D0913918

DICTIONARY OF

CLASSICAL
AND
THEORETICAL
MATHEMATICS

COMPREHENSIVE DICTIONARY
OF MATHEMATICS

Douglas N. Clark
Editor-in-Chief

Stan Gibilisco
Editorial Advisor

PUBLISHED VOLUMES

Analysis, Calculus, and Differential Equations
Douglas N. Clark

Algebra, Arithmetic, and Trigonometry
Steven G. Krantz

Classical and Theoretical Mathematics
Catherine Cavagnaro and William T. Haight, II

FORTHCOMING VOLUMES

Applied Mathematics for Engineers and Scientists
Emma Previato

The Comprehensive Dictionary of Mathematics
Douglas N. Clark

A VOLUME IN THE
COMPREHENSIVE DICTIONARY
OF MATHEMATICS

DICTIONARY OF

CLASSICAL AND THEORETICAL MATHEMATICS

Edited by
Catherine Cavagnaro
William T. Haight, II

CRC Press
Boca Raton   London   New York   Washington, D.C.

Library of Congress Cataloging-in-Publication Data

Dictionary of classical and theoretical mathematics / edited by Cahterine Cavagnaro and William T. Haight II.
      p. cm. — (Comprehensive dictionary of mathematics)
     ISBN 1-58488-050-3 (alk. paper)
      1. Mathematics—Dictionaries. I. Cavagnaro, Cahterine. II. Haight, William T. III.
  Series.

QA5 .D4984 2001
510′.3—dc21
                                                     00-068007

Visit the CRC Press Web site at www.crcpress.com

Preface

The *Dictionary of Classical and Theoretical Mathematics,* one volume of the *Comprehensive Dictionary of Mathematics,* includes entries from the fields of geometry, logic, number theory, set theory, and topology. The authors who contributed their work to this volume are professional mathematicians, active in both teaching and research.

The goal in writing this dictionary has been to define each term rigorously, not to author a large and comprehensive survey text in mathematics. Though it has remained our purpose to make each definition self-contained, some definitions unavoidably depend on others, and a modicum of "definition chasing" is necessitated. We hope this is minimal.

The authors have attempted to extend the scope of this dictionary to the fringes of commonly accepted higher mathematics. Surely, some readers will regard an excluded term as being mistakenly overlooked, and an included term as one "not quite yet cooked" by years of use by a broad mathematical community. Such differences in taste cannot be circumnavigated, even by our well-intentioned and diligent authors. Mathematics is a living and breathing entity, changing daily, so a list of included terms may be regarded only as a snapshot in time.

We thank the authors who spent countless hours composing original definitions. In particular, the help of Dr. Steve Benson, Dr. William Harris, and Dr. Tamara Hummel was key in organizing the collection of terms. Our hope is that this dictionary becomes a valuable source for students, teachers, researchers, and professionals.

Catherine Cavagnaro
William T. Haight, II

CONTRIBUTORS

Curtis Bennett
Bowling Green State University
Bowling Green, Ohio

Steve Benson
University of New Hampshire
Durham, New Hampshire

Catherine Cavagnaro
University of the South
Sewanee, Tennessee

Minevra Cordero
Texas Tech University
Lubbock, Texas

Douglas E. Ensley
Shippensburg University
Shippensburg, Pennsylvania

William T. Haight, II
University of the South
Sewanee, Tennessee

William Harris
Georgetown College
Georgetown, Kentucky

Phil Hotchkiss
University of St. Thomas
St. Paul, Minnesota

Matthew G. Hudelson
Washington State University
Pullman, Washington

Tamara Hummel
Allegheny College
Meadville, Pennsylvania

Mark J. Johnson
Central College
Pella, Iowa

Paul Kapitza
Illinois Wesleyan University
Bloomington, Illinois

Krystyna Kuperberg
Auburn University
Auburn, Alabama

Thomas LaFramboise
Marietta College
Marietta, Ohio

Adam Lewenberg
University of Akron
Akron, Ohio

Elena Marchisotto
California State University
Northridge, California

Rick Miranda
Colorado State University
Fort Collins, Colorado

Emma Previato
Boston University
Boston, Massachusetts

V.V. Raman
Rochester Institute of Technology
Pittsford, New York

David A. Singer
Case Western Reserve University
Cleveland, Ohio

David Smead
Furman University
Greenville, South Carolina

Sam Smith
St. Joseph's University
Philadelphia, Pennsylvania

Vonn Walter
Allegheny College
Meadville, Pennsylvania

Jerome Wolbert
University of Michigan
Ann Arbor, Michigan

Olga Yiparaki
University of Arizona
Tucson, Arizona

A

Abelian category   An additive category $\mathcal{C}$, which satisfies the following conditions, for any morphism $f \in \mathrm{Hom}_{\mathcal{C}}(X, Y)$:

(i.) f has a kernel (a morphism $i \in \mathrm{Hom}_{\mathcal{C}}$ (X', X) such that $fi = 0$) and a co-kernel (a morphism $p \in \mathrm{Hom}_{\mathcal{C}}(Y, Y')$ such that $pf = 0$);

(ii.) f may be factored as the composition of an epic (onto morphism) followed by a monic (one-to-one morphism) and this factorization is unique up to equivalent choices for these morphisms;

(iii.) if f is a monic, then it is a kernel; if f is an epic, then it is a co-kernel.

See additive category.

Abel's summation identity   If $a(n)$ is an arithmetic function (a real or complex valued function defined on the natural numbers), define

$$A(x) = \begin{cases} 0 & \text{if } x < 1, \\ \sum_{n \leq x} a(n) & \text{if } x \geq 1. \end{cases}$$

If the function f is continuously differentiable on the interval $[w, x]$, then

$$\sum_{w < n \leq x} a(n) f(n) = A(x) f(x)$$
$$- A(w) f(w)$$
$$- \int_w^x A(t) f'(t) \, dt .$$

abscissa of absolute convergence   For the Dirichlet series $\sum_{n=1}^{\infty} \frac{f(n)}{n^s}$, the real number σ_a, if it exists, such that the series converges absolutely for all complex numbers $s = x + iy$ with $x > \sigma_a$ but not for any s so that $x < \sigma_a$. If the series converges absolutely for all s, then $\sigma_a = -\infty$ and if the series fails to converge absolutely for any s, then $\sigma_a = \infty$. The set $\{x + iy : x > \sigma_a\}$ is called the *half plane of absolute convergence* for the series. *See also* abscissa of convergence.

abscissa of convergence   For the Dirichlet series $\sum_{n=1}^{\infty} \frac{f(n)}{n^s}$, the real number σ_c, if it exists, such that the series converges for all complex numbers $s = x + iy$ with $x > \sigma_c$ but not for any s so that $x < \sigma_c$. If the series converges absolutely for all s, then $\sigma_c = -\infty$ and if the series fails to converge absolutely for any s, then $\sigma_c = \infty$. The *abscissa of convergence* of the series is always less than or equal to the abscissa of absolute convergence ($\sigma_c \leq \sigma_a$). The set $\{x + iy : x > \sigma_c\}$ is called the *half plane of convergence* for the series. *See also* abscissa of absolute convergence.

absolute neighborhood retract   A topological space W such that, whenever (X, A) is a pair consisting of a (Hausdorff) normal space X and a closed subspace A, then any continuous function $f : A \longrightarrow W$ can be extended to a continuous function $F : U \longrightarrow W$, for U some open subset of X containing A. Any absolute retract is an *absolute neighborhood retract* (ANR). Another example of an ANR is the n-dimensional sphere, which is not an absolute retract.

absolute retract   A topological space W such that, whenever (X, A) is a pair consisting of a (Hausdorff) normal space X and a closed subspace A, then any continuous function $f : A \longrightarrow W$ can be extended to a continuous function $F : X \longrightarrow W$. For example, the unit interval is an *absolute retract;* this is the content of the Tietze Extension Theorem. *See also* absolute neighborhood retract.

absolute value   (1) If r is a real number, the quantity

$$|r| = \begin{cases} r & \text{if } r \geq 0, \\ -r & \text{if } r < 0. \end{cases}$$

Equivalently, $|r| = \sqrt{r^2}$. For example, $|-7| = |7| = 7$ and $|-1.237| = 1.237$. Also called *magnitude* of r.

(2) If $z = x + iy$ is a complex number, then $|z|$, also referred to as the *norm* or *modulus* of z, equals $\sqrt{x^2 + y^2}$. For example, $|1 - 2i| = \sqrt{1^2 + 2^2} = \sqrt{5}$.

(3) In $\mathbf{R}^n$ (Euclidean n space), the *absolute value* of an element is its (Euclidean) distance

1-58488-050-3/01/$0.00+$.50
© 2001 by CRC Press LLC

to the origin. That is,

$$|(a_1, a_2, \ldots, a_n)| = \sqrt{a_1^2 + a_2^2 + \cdots + a_n^2} \, .$$

In particular, if a is a real or complex number, then $|a|$ is the distance from a to 0.

abundant number   A positive integer n having the property that the sum of its positive divisors is greater than $2n$, i.e., $\sigma(n) > 2n$. For example, 24 is abundant, since

$$1 + 2 + 3 + 4 + 6 + 8 + 12 + 24 = 60 > 48 \, .$$

The smallest odd *abundant number* is 945. *Compare with* deficient number, perfect number.

accumulation point   A point x in a topological space X such that every neighborhood of x contains a point of X other than x. That is, for all open $U \subseteq X$ with $x \in U$, there is a $y \in U$ which is different from x. Equivalently, $x \in \overline{X \setminus \{x\}}$.

More generally, x is an *accumulation point* of a subset $A \subseteq X$ if every neighborhood of x contains a point of A other than x. That is, for all open $U \subseteq X$ with $x \in U$, there is a $y \in U \cap A$ which is different from x. Equivalently, $x \in \overline{A \setminus \{x\}}$.

additive category   A category $\mathcal{C}$ with the following properties:

(i.)   the Cartesian product of any two elements of $\mathrm{Obj}(\mathcal{C})$ is again in $\mathrm{Obj}(\mathcal{C})$;

(ii.) $\mathrm{Hom}_{\mathcal{C}}(A, B)$ is an additive Abelian group with identity element 0, for any $A, B \in \mathrm{Obj}(\mathcal{C})$;

(iii.)   the distributive laws $f(g_1 + g_2) = fg_1 + fg_1$ and $(f_1 + f_2)g = f_1 g + f_2 g$ hold for morphisms when the compositions are defined. *See* category.

additive function   An arithmetic function f having the property that $f(mn) = f(m) + f(n)$ whenever m and n are relatively prime. (*See* arithmetic function). For example, ω, the number of distinct prime divisors function, is additive. The values of an *additive function* depend only on its values at powers of primes: if $n = p_1^{i_1} \cdots p_k^{i_k}$ and f is additive, then $f(n) = f(p_1^{i_1}) + \ldots + f(p_k^{i_k})$. *See also* completely additive function.

additive functor   An additive functor $F : \mathcal{C} \to \mathcal{D}$, between two additive categories, such that $F(f + g) = F(f) + F(g)$ for any $f, g \in \mathrm{Hom}_{\mathcal{C}}(A, B)$. *See* additive category, functor.

Adem relations   The relations in the Steenrod algebra which describe a product of pth power or square operations as a linear combination of products of these operations. For the square operations ($p = 2$), when $0 < i < 2j$,

$$Sq^i \, Sq^j = \sum_{0 \leq k \leq [i/2]} \binom{j - k - 1}{i - 2k} Sq^{i+j-k} \, Sq^k,$$

where $[i/2]$ is the greatest integer less than or equal to $i/2$ and the binomial coefficients in the sum are taken mod 2, since the square operations are a $\mathbf{Z}/2$-algebra.

As a consequence of the values of the binomial coefficients, $Sq^{2n-1} \, Sq^n = 0$ for all values of n.

The relations for Steenrod algebra of pth power operations are similar.

adjoint functor   If X is a fixed object in a category $\mathcal{X}$, the covariant functor $\mathrm{Hom}_* : \mathcal{X} \to \mathbf{Sets}$ maps $A \in \mathrm{Obj}(\mathcal{X})$ to $\mathrm{Hom}_{\mathcal{X}}(X, A)$; $f \in \mathrm{Hom}_{\mathcal{X}}(A, A')$ is mapped to $f_* : \mathrm{Hom}_{\mathcal{X}}(X, A) \to \mathrm{Hom}_{\mathcal{X}}(X, A')$ by $g \mapsto fg$. The contravariant functor $\mathrm{Hom}^* : \mathcal{X} \to \mathbf{Sets}$ maps $A \in \mathrm{Obj}(\mathcal{X})$ to $\mathrm{Hom}_{\mathcal{X}}(A, X)$; $f \in \mathrm{Hom}_{\mathcal{X}}(A, A')$ is mapped to

$$f^* : \mathrm{Hom}_{\mathcal{X}}(A', X) \to \mathrm{Hom}_{\mathcal{X}}(A, X) \, ,$$

by $g \mapsto gf$.

Let $\mathcal{C}, \mathcal{D}$ be categories. Two covariant functors $F : \mathcal{C} \to \mathcal{D}$ and $G : \mathcal{D} \to \mathcal{C}$ are *adjoint functors* if, for any $A, A' \in \mathrm{Obj}(\mathcal{C})$, $B, B' \in \mathrm{Obj}(\mathcal{D})$, there exists a bijection

$$\phi : \mathrm{Hom}_{\mathcal{C}}(A, G(B)) \to \mathrm{Hom}_{\mathcal{D}}(F(A), B)$$

that makes the following diagrams commute for any $f : A \to A'$ in $\mathcal{C}$, $g : B \to B'$ in $\mathcal{D}$:

$$\text{Hom}_{\mathcal{C}}(A, G(B)) \xrightarrow{f^*} \text{Hom}_{\mathcal{C}}(A', G(B))$$
$$\phi\downarrow \qquad\qquad\qquad \phi\downarrow$$
$$\text{Hom}_{\mathcal{D}}(F(A), B) \xrightarrow{(F(f))^*} \text{Hom}_{\mathcal{D}}(F(A'), B)$$

$$\text{Hom}_{\mathcal{C}}(A, G(B)) \xrightarrow{(G(g))_*} \text{Hom}_{\mathcal{C}}(A, G(B'))$$
$$\phi\downarrow \qquad\qquad\qquad \phi\downarrow$$
$$\text{Hom}_{\mathcal{D}}(F(A), B) \xrightarrow{g_*} \text{Hom}_{\mathcal{D}}(F(A), B')$$

See category of sets.

alephs   Form the sequence of infinite cardinal numbers ($\aleph_\alpha$), where α is an ordinal number.

Alexander's Horned Sphere   An example of a two sphere in $\mathbf{R}^3$ whose complement in $\mathbf{R}^3$ is not topologically equivalent to the complement of the standard two sphere $S^2 \subset \mathbf{R}^3$.

This space may be constructed as follows: On the standard two sphere S^2, choose two mutually disjoint disks and extend each to form two "horns" whose tips form a pair of parallel disks. On each of the parallel disks, form a pair of horns with parallel disk tips in which each pair of horns interlocks the other and where the distance between each pair of horn tips is half the previous distance. Continuing this process, at stage n, 2^n pairwise linked horns are created.

In the limit, as the number of stages of the construction approaches infinity, the tips of the horns form a set of limit points in $\mathbf{R}^3$ homeomorphic to the Cantor set. The resulting surface is homeomorphic to the standard two sphere S^2 but the complement in $\mathbf{R}^3$ is not simply connected.

algebra of sets   A collection of subsets $\mathcal{S}$ of a non-empty set X which contains X and is closed with respect to the formation of finite unions, intersections, and differences. More precisely,

(i.) $X \in \mathcal{S}$;

(ii.) if $A, B \in \mathcal{S}$, then $A \cup B, A \cap B$, and $A \backslash B$ are also in $\mathcal{S}$.

See union, difference of sets.

algebraic number   (**1**) A complex number which is a zero of a polynomial with rational coefficients (i.e., α is *algebraic* if there exist ratio-

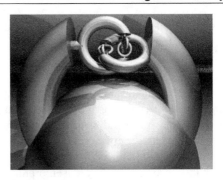

Alexander's Horned Sphere. Graphic rendered by PovRay.

nal numbers $a_0, a_1, \ldots, a_n$ so that $\sum_{i=0}^{n} a_i\alpha^i = 0$).

For example, $\sqrt{2}$ is an *algebraic number* since it satisfies the equation $x^2 - 2 = 0$. Since there is no polynomial $p(x)$ with rational coefficients such that $p(\pi) = 0$, we see that π is *not* an algebraic number. A complex number that is not an algebraic number is called a *transcendental number*.

(**2**) If F is a field, then α is said to be *algebraic over F* if α is a zero of a polynomial having coefficients in F. That is, if there exist elements $f_0, f_1, f_2, \ldots, f_n$ of F so that $f_0 + f_1\alpha + f_2\alpha^2 \cdots + f_n\alpha^n = 0$, then α is algebraic over F.

algebraic number field   A subfield of the complex numbers consisting entirely of algebraic numbers. *See also* algebraic number.

algebraic number theory   That branch of mathematics involving the study of algebraic numbers and their generalizations. It can be argued that the genesis of *algebraic number theory* was Fermat's Last Theorem since much of the results and techniques of the subject sprung directly or indirectly from attempts to prove the Fermat conjecture.

algebraic variety   Let A be a polynomial ring $k[x_1, \ldots, x_n]$ over a field k. An *affine algebraic variety* is a closed subset of A^n (in the Zariski topology of A^n) which is not the union of two proper (Zariski) closed subsets of A^n. In the Zariski topology, a closed set is the set of common zeros of a set of polynomials. Thus, an affine algebraic variety is a subset of A^n which is the set of common zeros of a set of polynomials but which cannot be expressed as the union of two such sets.

The topology on an affine variety is inherited from A^n.

In general, an (abstract) algebraic variety is a topological space with open sets U_i whose union is the whole space and each of which has an affine algebraic variety structure so that the induced variety structures (from U_i and U_j) on each intersection $U_i \cap U_j$ are isomorphic.

The solutions to any polynomial equation form an algebraic variety. Real and complex projective spaces can be described as algebraic varieties (k is the field of real or complex numbers, respectively).

altitude In plane geometry, a line segment joining a vertex of a triangle to the line through the opposite side and perpendicular to the line. The term is also used to describe the length of the line segment. The area of a triangle is given by one half the product of the length of any side and the length of the corresponding *altitude*.

amicable pair of integers Two positive integers m and n such that the sum of the positive divisors of both m and n is equal to the sum of m and n, i.e., $\sigma(m) = \sigma(n) = m + n$. For example, 220 and 284 form an amicable pair, since

$$\sigma(220) = \sigma(284) = 504 .$$

A perfect number forms an amicable pair with itself.

analytic number theory That branch of mathematics in which the methods and ideas of real and complex analysis are applied to problems concerning integers.

analytic set The continuous image of a Borel set. More precisely, if X is a Polish space and $A \subseteq X$, then A is analytic if there is a Borel set B contained in a Polish space Y and a continuous $f : X \to Y$ with $f(A) = B$. Equivalently, A is analytic if it is the projection in X of a closed set

$$C \subseteq X \times \mathbf{N}^\mathbf{N} ,$$

where $\mathbf{N}^\mathbf{N}$ is the Baire space. Every Borel set is analytic, but there are analytic sets that are not Borel. The collection of analytic sets is denoted Σ_1^1. *See also* Borel set, projective set.

annulus A topological space homeomorphic to the product of the sphere S^n and the closed unit interval I. The term sometimes refers specifically to a closed subset of the plane bounded by two concentric circles.

antichain A subset A of a partially ordered set $(P, \leq)$ such that any two distinct elements $x, y \in A$ are not comparable under the ordering $\leq$. Symbolically, neither $x \leq y$ nor $y \leq x$ for any $x, y \in A$.

arc A subset of a topological space, homeomorphic to the closed unit interval $[0, 1]$.

arcwise connected component If p is a point in a topological space X, then the *arcwise connected component* of p in X is the set of points q in X such that there is an arc (in X) joining p to q. That is, for any point q distinct from p in the arc component of p there is a homeomorphism $\phi : [0, 1] \longrightarrow J$ of the unit interval onto some subspace J containing p and q. The arcwise connected component of p is the largest arcwise connected subspace of X containing p.

arcwise connected topological space A topological space X such that, given any two distinct points p and q in X, there is a subspace J of X homeomorphic to the unit interval $[0, 1]$ containing both p and q.

arithmetical hierarchy A method of classifying the complexity of a set of natural numbers based on the quantifier complexity of its definition. The *arithmetical hierarchy* consists of classes of sets Σ_n^0, Π_n^0, and Δ_n^0, for $n \geq 0$.

A set A is in $\Sigma_0^0 = \Pi_0^0$ if it is recursive (computable). For $n \geq 1$, a set A is in Σ_n^0 if there is a computable (recursive) $(n + 1)$–ary relation R such that for all natural numbers x,

$$x \in A \iff (\exists y_1)(\forall y_2) \ldots (Q_n y_n) R(x, \overline{y}),$$

where Q_n is $\exists$ if n is odd and Q_n is $\forall$ if n is odd, and where $\overline{y}$ abbreviates $y_1, \ldots, y_n$. For $n \geq 1$, a set A is in Π_n^0 if there is a computable (recursive) $(n + 1)$–ary relation R such that for all natural numbers x,

$$x \in A \iff (\forall y_1)(\exists y_2)\ldots(Q_n y_n) R(x, \overline{y}),$$

where Q_n is $\exists$ if n is even and Q_n is $\forall$ if n is odd. For $n \geq 0$, a set A is in Δ_n^0 if it is in both Σ_n^0 and Π_n^0.

Note that it suffices to define the classes Σ_n^0 and Π_n^0 as above since, using a computable coding function, pairs of like quantifiers (for example, $(\exists y_1)(\exists y_2)$) can be contracted to a single quantifier $((\exists y))$. The superscript 0 in Σ_n^0, Π_n^0, Δ_n^0 is sometimes omitted and indicates classes in the arithmetical hierarchy, as opposed to the analytical hierarchy.

A set A is *arithmetical* if it belongs to the arithmetical hierarchy; i.e., if, for some n, A is in Σ_n^0 or Π_n^0. For example, any computably (recursively) enumerable set is in Σ_1^0.

arithmetical set   A set A which belongs to the arithmetical hierarchy; i.e., for some n, A is in Σ_n^0 or Π_n^0. *See* arithmetical hierarchy. For example, any computably (recursively) enumerable set is in Σ_1^0.

arithmetic function   A function whose domain is the set of positive integers. Usually, an *arithmetic function* measures some property of an integer, e.g., the Euler phi function ϕ or the sum of divisors function σ. The properties of the function itself, such as its order of growth or whether or not it is multiplicative, are often studied. Arithmetic functions are also called number theoretic functions.

Aronszajn tree   A tree of height ω_1 which has no uncountable branches or levels. Thus, for each $\alpha < \omega_1$, the α-level of T, $\mathrm{Lev}_\alpha(T)$, given by

$$\{t \in T : \text{ordertype}(\{s \in T : s < t\}) = \alpha\}$$

is countable, $\mathrm{Lev}_{\omega_1}(T)$ is the first empty level of T, and any set $B \subseteq T$ which is totally ordered by $<$ (branch) is countable. An *Aronszajn tree* is constructible in ZFC without any extra set-theoretic hypotheses.

For any regular cardinal κ, a κ-*Aronszajn tree* is a tree of height κ in which all levels have size less than κ and all branches have length less than κ. *See also* Suslin tree, Kurepa tree.

associated fiber bundle   A concept in the theory of fiber bundles. A fiber bundle ζ consists of a space B called the base space, a space E called the total space, a space F called the fiber, a topological group G of transformations of F, and a map $\pi : E \longrightarrow B$. There is a covering of B by open sets U_i and homeomorphisms $\phi_i : U_i \times F \longrightarrow E_i = \pi^{-1}(U_i)$ such that $\pi \circ \phi_i(x, V) = x$. This identifies $\pi^{-1}(x)$ with the fiber F. When two sets U_i and U_j overlap, the two identifications are related by coordinate transformations $g_{ij}(x)$ of F, which are required to be continuously varying elements of G. If G also acts as a group of transformations on a space F', then the *associated fiber bundle* $\zeta' = \pi' : E' \longrightarrow B$ is the (uniquely determined) fiber bundle with the same base space B, fiber F', and the same coordinate transformations as ζ.

associated principal fiber bundle   The associated fiber bundle, of a fiber bundle ζ, with the fiber F replaced by the group G. *See* associated fiber bundle. The group acts by left multiplication, and the coordinate transformations g_{ij} are the same as those of the bundle ζ.

atomic formula   Let $\mathcal{L}$ be a first order language. An *atomic formula* is an expression which has the form $P(t_1, \ldots, t_n)$, where P is an n-place predicate symbol of $\mathcal{L}$ and $t_1, \ldots, t_n$ are terms of $\mathcal{L}$. If $\mathcal{L}$ contains equality ($=$), then $=$ is viewed as a two-place predicate. Consequently, if t_1 and t_2 are terms, then $t_1 = t_2$ is an atomic formula.

atomic model   A model A in a language L such that every n-tuple of elements of A satisfies a complete formula in T, the theory of A. That is, for any $\overline{a} \in A^n$, there is an L-formula $\theta(\overline{x})$ such that $A \models \theta(\overline{a})$, and for any L-formula ϕ, either $T \vdash \forall \overline{x}(\theta(\overline{x}) \rightarrow \phi(\overline{x}))$ or $T \vdash \forall \overline{x}(\theta(\overline{x}) \rightarrow \neg\phi(\overline{x}))$. This is equivalent to the complete type of every $\overline{a}$ being principal. Any finite model is atomic, as is the standard model of number theory.

atom of a Boolean algebra   If $(\mathcal{B}, \vee, \wedge, \sim, 1, 0)$ is a Boolean algebra, $a \in \mathcal{B}$ is an atom if it is a minimal element of $\mathcal{B}\backslash\{0\}$. For example, in the Boolean algebra of the power set of any nonempty set, any singleton set is an atom.

automorphism   Let $\mathcal{L}$ be a first order language and let $\mathcal{A}$ be a structure for $\mathcal{L}$. An *automorphism* of $\mathcal{A}$ is an isomorphism from $\mathcal{A}$ onto itself. *See* isomorphism.

axiomatic set theory   A collection of statements concerning set theory which can be proved from a collection of fundamental axioms. The validity of the statements in the theory plays no role; rather, one is only concerned with the fact that they can be deduced from the axioms.

Axiom of Choice   Suppose that $\{X_\alpha\}_{\alpha \in \Gamma}$ is a family of non-empty, pairwise disjoint sets. Then there exists a set Y which consists of exactly one element from each set in the family. Equivalently, given any family of non-empty sets $\{X_\alpha\}_{\alpha \in \Gamma}$, there exists a function $f : \{X_\alpha\}_{\alpha \in \Gamma} \to \bigcup_{\alpha \in \Lambda} X_\alpha$ such that $f(X_\alpha) \in X_\alpha$ for each $\alpha \in \Gamma$.

The existence of such a set Y or function f can be proved from the Zermelo-Fraenkel axioms when there are only finitely many sets in the family. However, when there are infinitely many sets in the family it is impossible to prove that such Y, f exist or do not exist. Therefore, neither the *Axiom of Choice* nor its negation can be proved from the axioms of Zermelo-Fraenkel set theory.

Axiom of Comprehension   Also called Axiom of Separation. *See* Axiom of Separation.

Axiom of Constructibility   Every set is constructible. *See* constructible set.

Axiom of Dependent Choice   *See* principle of dependent choices.

Axiom of Determinancy   For any set $X \subseteq \omega^\omega$, the game G_X is determined. This axiom contradicts the Axiom of Choice. *See* determined.

Axiom of Equality   If two sets are equal, then they have the same elements. This is the converse of the Axiom of Extensionality and is considered to be an axiom of logic, not an axiom of set theory.

Axiom of Extensionality   If two sets have the same elements, then they are equal. This is one of the axioms of Zermelo-Fraenkel set theory.

Axiom of Foundation   Same as the Axiom of Regularity. *See* Axiom of Regularity.

Axiom of Infinity   There exists an infinite set. This is one of the axioms of Zermelo-Fraenkel set theory. *See* infinite set.

Axiom of Regularity   Every non-empty set has an $\in$-minimal element. More precisely, every non-empty set S contains an element $x \in S$ with the property that there is no element $y \in S$ such that $y \in x$. This is one of the axioms of Zermelo-Fraenkel set theory.

Axiom of Replacement   If f is a function, then, for every set X, there exists a set $f(X) = \{f(x) : x \in X\}$. This is one of the axioms of Zermelo-Fraenkel set theory.

Axiom of Separation   If P is a property and X is a set, then there exists a set $Y = \{x \in X : x$ satisfies property $P\}$.

This is one of the axioms of Zermelo-Fraenkel set theory. It is a weaker version of the Axiom of Comprehension: if P is a property, then there exists a set $Y = \{X : X$ satisfies property $P\}$. Russell's Paradox shows that the Axiom of Comprehension is false for sets. *See also* Russell's Paradox.

Axiom of Subsets   Same as the Axiom of Separation. *See* Axiom of Separation.

Axiom of the Empty Set   There exists a set $\emptyset$ which has no elements.

Axiom of the Power Set   For every set X, there exists a set $P(X)$, the set of all subsets of X. This is one of the axioms of Zermelo-Fraenkel set theory.

Axiom of the Unordered Pair   If X and Y are sets, then there exists a set $\{X, Y\}$. This axiom, also known as the Axiom of Pairing, is one of the axioms of Zermelo-Fraenkel set theory.

Axiom of Union   For any set S, there exists a set that is the union of all the elements of S.

B

Baire class   The Baire classes B_α are an increasing sequence of families of functions defined inductively for $\alpha < \omega_1$. B_0 is the set of continuous functions. For $\alpha > 0$, f is in Baire class α if there is a sequence of functions $\{f_n\}$ converging pointwise to f, with $f_n \in B_{\beta_n}$ and $\beta_n < \alpha$ for each n. Thus, f is in Baire class 1 (or is Baire-1) if it is the pointwise limit of a sequence of continuous functions. In some cases, it is useful to define the classes so that if $f \in B_\alpha$, then $f \notin B_\beta$ for any $\beta < \alpha$. *See also* Baire function.

Baire function   A function belonging to one of the Baire classes, B_α, for some $\alpha < \omega_1$. Equivalently, the set of *Baire functions* in a topological space is the smallest collection containing all continuous functions which is closed under pointwise limits. *See* Baire class.

It is a theorem that f is a Baire function if and only if f is Borel measurable, that is, if and only if $f^{-1}(U)$ is a Borel set for any open set U.

Baire measurable function   A function $f : X \to Y$, where X and Y are topological spaces, such that the inverse image of any open set has the Baire property. *See* Baire property. That is, if $V \subseteq Y$ is open, then

$$f^{-1}(V) = U \Delta C = (U \setminus C) \cup (C \setminus U),$$

where $U \subseteq X$ is open and $C \subseteq X$ is meager.

Baire property   A set that can be written as an open set modulo a first category or meager set. That is, X has the *Baire property* if there is an open set U and a meager set C with

$$X = U \Delta C = (U \setminus C) \cup (C \setminus U).$$

Since the meager sets form a σ-ideal, this happens if and only if there is an open set U and meager sets C and D with $X = (U \setminus C) \cup D$. Every Borel set has the Baire property; in fact, every analytic set has the Baire property.

Baire space   (1) A topological space X such that no nonempty open set in X is meager (first category). That is, no open set $U \neq \emptyset$ in X may be written as a countable union of nowhere dense sets. Equivalently, X is a *Baire space* if and only if the intersection of any countable collection of dense open sets in X is dense, which is true if and only if, for any countable collection of closed sets $\{C_n\}$ with empty interior, their union $\cup C_n$ also has empty interior. The Baire Category Theorem states that any complete metric space is a Baire space.

(2) The Baire space is the set of all infinite sequences of natural numbers, $\mathbf{N}^\mathbf{N}$, with the product topology and using the discrete topology on each copy of $\mathbf{N}$. Thus, U is a basic open set in $\mathbf{N}^\mathbf{N}$ if there is a finite sequence of natural numbers σ such that U is the set of all infinite sequences which begin with σ. The Baire space is homeomorphic to the irrationals.

bar construction   For a group G, one can construct a space BG as the geometric realization of the following simplicial complex. The faces F_n in simplicial degree n are given by $(n + 1)$-tuples of elements of G. The boundary maps $F_n \longrightarrow F_{n-1}$ are given by the simplicial boundary formula

$$\sum_{i=0}^{n} (-1)^i (g_0, \dots, \hat{g}_i, \dots, g_n)$$

where the notation $\hat{g}_i$ indicates that g_i is omitted to obtain an n-tuple. The ith degeneracy map $s_i : F_n \longrightarrow F_{n+1}$ is given by inserting the group identity element in the ith position.

Example: $B(\mathbf{Z}/2)$, the classifying space of the group $\mathbf{Z}/2$, is RP^∞, real infinite projective space (the union of RP^n for all n positive integers).

The *bar construction* has many generalizations and is a useful means of constructing the nerve of a category or the classifying space of a group, which determines the vector bundles of a manifold with the group acting on the fiber.

base of number system   The number b, in use, when a real number r is written in the form

$$r = \sum_{j=-\infty}^{N} r_j b^j,$$

where each $r_j = 0, 1, ..., b - 1$, and r is represented in the notation

$$r = r_N r_{N-1} \cdots r_0 . r_{-1} r_{-2} \cdots .$$

For example, the base of the standard decimal system is 10 and we need the digits 0, 1, 2, 3, 4, 5, 6, 7, 8, and 9 in order to use this system. Similarly, we use only the digits 0 and 1 in the binary system; this is a "base 2" system. In the base b system, the number 10215.2011 is equivalent to the decimal number

$$1 \times b^4 + 0 \times b^3 + 2 \times b^2 + 1 \times b + 5 + 2 \times b^{-1}$$

$$+ 0 \times b^{-2} + 1 \times b^{-3} + 1 \times b^{-4} .$$

That is, each place represents a specific power of the base b. *See also* radix.

Bernays-Gödel set theory   An axiomatic set theory, which is based on axioms other than those of Zermelo-Fraenkel set theory. *Bernays-Gödel set theory* considers two types of objects: sets and classes. Every set is a class, but the converse is not true; classes that are not sets are called proper classes. This theory has the Axioms of Infinity, Union, Power Set, Replacement, Regularity, and Unordered Pair for sets from Zermelo-Fraenkel set theory. It also has the following axioms, with classes written in :

(i.) Axiom of Extensionality (for classes): Suppose that $\mathbf{X}$ and $\mathbf{Y}$ are two classes such that $U \in \mathbf{X}$ if and only if $U \in \mathbf{Y}$ for all set U. Then $\mathbf{X} = \mathbf{Y}$.

(ii.) If $\mathbf{X} \in \mathbf{Y}$, then $\mathbf{X}$ is a set.

(iii.) Axiom of Comprehension: For any formula $F(X)$ having sets as variables there exists a class $\mathbf{Y}$ consisting of all sets satisfying the formula $F(X)$.

Bertrand's postulate   If x is a real number greater than 1, then there is at least one prime number p so that $x < p < 2x$. *Bertrand's Postulate* was conjectured to be true by the French mathematician Joseph Louis Francois Bertrand and later proved by the Russian mathematician Pafnuty Lvovich Tchebychef.

Betti number   Suppose X is a space whose homology groups are finitely generated. Then the kth homology group is isomorphic to the direct sum of a torsion group T_k and a free Abelian group B_k. The kth *Betti number* $b_k(X)$ of X is the rank of B_k. Equivalently, $b_k(X)$ is the dimension of $H_k(X, Q)$, the kth homology group with rational coefficients, viewed as a vector space over the rationals. For example, $b_0(X)$ is the number of connected components of X.

bijection   A function $f : X \to Y$, between two sets, with the following two properties:

(i.) f is one-to-one (if $x_1, x_2 \in X$ and $f(x_1) = f(x_2)$, then $x_1 = x_2$);

(ii.) f is onto (for any $y \in Y$ there exists an $x \in X$ such that $f(x) = y$).

See function.

binomial coefficient   (1) If n and k are nonnegative integers with $k \leq n$, then the *binomial coefficient* $\binom{n}{k}$ equals $\frac{n!}{k!(n-k)!}$.

(2) The binomial coefficient $\binom{n}{k}$ also represents the number of ways to choose k distinct items from among n distinct items, without regard to the order of choosing.

(3) The binomial coefficient $\binom{n}{k}$ is the kth entry in the nth row of Pascal's Triangle. It must be noted that Pascal's Triangle begins with row 0, and each row begins with entry 0. *See* Pascal's triangle.

Binomial Theorem   If a and b are elements of a commutative ring and n is a non-negative integer, then $(a + b)^n = \sum_{k=0}^{n} \binom{n}{k} a^k b^{n-k}$, where $\binom{n}{k}$ is the binomial coefficient. *See* binomial coefficient.

Bockstein operation   In cohomology theory, a cohomology operation is a natural transformation between two cohomology functors. If $0 \to A \to B \to C \to 0$ is a short exact sequence of modules over a ring R, and if $X \subset Y$ are topological spaces, then there is a long exact sequence in cohomology:

$$\cdots \to H^q(X, Y; A) \to H^q(X, Y; B) \to$$

$$H^q(X, Y; C) \to$$

$$H^{q+1}(X, Y; A) \to H^{q+1}(X, Y; B) \to \cdots .$$

The homomorphism

$$\beta : H^q(X, Y; C) \to H^{q+1}(X, Y; A)$$

is the *Bockstein (cohomology) operation*.

Bolzano-Weierstrass Theorem Every bounded sequence in $\mathbf{R}$ has a convergent subsequence. That is, if

$$\{x_n : n \in \mathbf{N}\} \subseteq [\mathbf{a}, \mathbf{b}]$$

is an infinite sequence, then there is an increasing sequence $\{n_k : k \in \mathbf{N}\} \subseteq \mathbf{N}$ such that $\{x_{n_k} : k \in \mathbf{N}\}$ converges.

Boolean algebra A non-empty set X, along with two binary operations $\cup$ and $\cap$ (called union and intersection, respectively), a unary operation $'$ (called complement), and two elements $0, 1 \in X$ which satisfy the following properties for all $A, B, C \in X$.

(i.) $A \cup (B \cup C) = (A \cup B) \cup C$
(ii.) $A \cap (B \cap C) = (A \cap B) \cap C$
(iii.) $A \cup B = B \cup A$
(iv.) $A \cap B = B \cap A$
(v.) $A \cap (B \cup C) = (A \cap B) \cup (A \cap C)$
(vi.) $A \cup (B \cap C) = (A \cup B) \cap (A \cup C)$
(vii.) $A \cup 0 = A$ and $A \cap 1 = A$
(viii.) There exists an element A' so that $A \cup A' = 1$ and $A \cap A' = 0$.

Borel measurable function A function $f : X \to Y$, for X, Y topological spaces, such that the inverse image of any open set is a Borel set. This is equivalent to requiring the inverse image of any Borel set to be Borel. Any continuous function is Borel measurable.

It is a theorem that f is Borel measurable if and only if f is a Baire function. *See* Baire function.

Borel set The collection $\mathcal{B}$ of *Borel sets* of a topological space X is the smallest σ-algebra containing all open sets of X. That is, in addition to containing open sets, $\mathcal{B}$ must be closed under complements and countable intersections (and, thus, is also closed under countable unions). For comparison, the topology on X is closed under arbitrary unions but only finite intersections.

Borel sets may also be defined inductively: let Σ_1^0 denote the collection of open sets and Π_1^0 the closed sets. Then for $1 < \alpha < \omega_1$, $A \in \Sigma_\alpha^0$ if and only if

$$A = \cup_{n \in \mathbf{N}} A_n$$

where, for each $n \in \mathbf{N}$, $A_n \in \Pi_{\alpha_n}^0$ and $\alpha_n < \alpha$. A set B is in Π_α^0 if and only if the complement

of B is in Σ_α^0. Then the collection of all Borel sets is

$$\mathcal{B} = \cup_{\alpha < \omega_1} \Sigma_\alpha^0 = \cup_{\alpha < \omega_1} \Pi_\alpha^0 .$$

Sets in Σ_2^0 are also known as F_σ sets; sets in Π_2^0 are G_δ.

If the space X is metrizable, then closed sets are G_δ and open sets are F_σ. In this case, we have for all $\alpha < \omega_1$,

$$\Sigma_\alpha^0 \cup \Pi_\alpha^0 \subseteq \Sigma_{\alpha+1}^0 \cap \Pi_{\alpha+1}^0 .$$

This puts the Borel sets in a hierarchy of length ω_1 known as the Borel hierarchy. *See also* projective set.

bound (1) An upper *bound* on a set, S, of real numbers is a number u so that $u \geq s$ for all $s \in S$. If such a u exists, S is said to be *bounded above* by u. Note that if u is an upper bound for the set S, then so is any number larger than u. *See also* least upper bound.

(2) A lower bound on a set, S, of real numbers is a number ℓ so that $\ell \leq s$ for all $s \in S$. If such an ℓ exists, S is said to be *bounded below* by ℓ. Note that if ℓ is a lower bound for the set S, then so is any number smaller than ℓ. *See* greatest lower bound.

(3) A bound on a set, S, of real numbers is a number b so that $|s| \leq b$ for all $s \in S$.

boundary group (homology) If $\{C_n, \partial_n\}$ is a chain complex (of Abelian groups), then the kth *boundary group* B_k is the subgroup of C_k consisting of elements of the form ∂c for c in C_{k+1}. That is, $B_k = \partial C_{k+1}$.

boundary operator A chain complex $\{C_n, \partial_n\}$ consists of a sequence of groups or modules over a ring R, together with homomorphisms $\partial_n : C_n \longrightarrow C_{n-1}$, such that $\partial_{n-1} \circ \partial_n = 0$. The homomorphisms ∂_n are called the *boundary operators*. Specifically, if K is an ordered simplicial complex and C_n is the free Abelian group generated by the n-dimensional simplices, then the boundary operator is defined by taking any n-simplex σ to the alternating sum of its $n-1$-dimensional faces. This definition is then extended to a homomorphism.

bounded quantifier The quantifiers $\forall x < y$ and $\exists x < y$. The statement $\forall x < y \, \phi(x)$ is equivalent to $\forall x(x < y \to \phi(x))$, and $\exists x < y\,\phi(x)$ is equivalent to $\exists x(x < y \land \phi(x))$.

More generally, $\forall x \in y\,\phi(x)$ is equivalent to $\forall x(x \in y \to \phi(x))$ and $\exists x \in y\,\phi(x)$ is equivalent to $\exists x(x \in y \land \phi(x))$.

bound variable   Let $\mathcal{L}$ be a first-order language and let φ be a well-formed formula of $\mathcal{L}$. An occurrence of a variable v in φ is *bound* if it occurs as the variable of a quantifier or within the scope of a quantifier $\forall v$ or $\exists v$. The scope of the quantifier $\forall v$ in a formula $\forall v\alpha$ is α.

For example, the first occurrence of the variable v_1 is free, while the remaining occurrences are bound in the formula

$$\forall v_2(v_1 = v_2 \to \forall v_1(v_1 = v_3)).$$

All occurrences of the variable v_1 are bound in the formula

$$\forall v_1(v_1 = v_2 \to \forall v_1(v_1 = v_3)).$$

box topology   A topology on the Cartesian product

$$\prod_{\alpha \in A} X_\alpha$$

of a collection of topological spaces X_α, having as a basis the set of all open boxes, $\prod_{\alpha \in A} U_\alpha$, where each U_α is an open subset of X_α. The difference between this and the product topology is that in the *box topology,* there are no restrictions on any of the U_α.

Brouwer Fixed-Point Theorem   Any continuous mapping f of a finite product of copies of $[0, 1]$ to itself, or of S^n to itself, has a fixed point, that is, a point z such that $f(z) = z$.

Intuitively, if a piece of paper is taken off a table, crumpled up, and laid back down on the same part of the table, then at least one point is exactly above the same point on the table that it was originally.

bundle group   A group that acts (continuously) on a vector bundle or fiber bundle $E \longrightarrow B$ and preserves fibers (so the action restricts to an action on each inverse image of a point in B). For example, the real orthogonal group $O(n)$ is a *bundle group* for any rank n real vector bundle. If the bundle is orientable, then $SO(n)$ is also a bundle group for the vector bundle.

The bundle group may also be called the *structure group* of the bundle.

bundle mapping   A fiber preserving map $g : E \longrightarrow E'$, where $p : E \longrightarrow B$ and $p' : E' \longrightarrow B'$ are fiber bundles. If the bundles are smooth vector bundles, then g must be a smooth map and linear on the vector space fibers.

Example: When a manifold is embedded in $\mathbf{R}^n$, it has both a tangent and a normal bundle. The direct sum of these is the trivial bundle $M \times \mathbf{R}^n$; each inclusion into the trivial rank n bundle is a *bundle mapping.*

bundle of planes   A fiber bundle whose fibers are all homeomorphic to $\mathbf{R}^2$. A canonical example of this is given by considering the Grassmann manifold of planes in $\mathbf{R}^n$. Each point corresponds to a plane in $\mathbf{R}^n$ in the same way each point of the projective space $\mathbf{RP}^{n-1}$ corresponds to a line in $\mathbf{R}^n$. The *bundle of planes* over this manifold is given by allowing the fiber over each point in the manifold to be the actual plane represented by that point. If one considers the manifold as the collection of names of the planes, then the bundle is the collection of planes, parameterized by their "names".

C

canonical bundle   If the points of a space represent (continuously parameterized) geometric objects, then the space has a *canonical bundle* given by setting the fiber above each point to be the geometric object to which that point corresponds. Examples include the canonical line bundle of projective space and the canonical vector bundle over a Grassmann manifold (the manifold of affine n-spaces in $\mathbf{R}^m$).

canonical line bundle   Projective space $\mathbf{RP}^n$ can be considered as the space of all lines in $\mathbf{R}^{n+1}$ which go through the origin or, equivalently, as the quotient of S^{n+1} formed by identifying each point with its negative. The *canonical line bundle* over $\mathbf{RP}^n$ is the rank one vector bundle formed by taking as fiber over a point in $\mathbf{RP}^n$ the actual line that the point represents.

Example: $\mathbf{RP}^1$ is homeomorphic to S^1; the canonical line bundle over $\mathbf{RP}^1$ is homeomorphic to the Möbius band.

There are also projective spaces formed over complex or quaternionic space, where a line is a complex or quaternionic line.

Cantor-Bernstein Theorem   If A and B are sets, and $f: A \to B$, $g: B \to A$ are injective functions, then there exists a bijection $h: A \to B$. This theorem is also known as the Cantor-Schröder-Bernstein Theorem or the Schröder-Bernstein Theorem.

Cantor-Schröder-Bernstein Theorem   *See* Cantor-Bernstein Theorem.

Cantor set   (1) (The *standard Cantor set.*) A subset of $\mathbf{R}^1$ which is an example of a totally disconnected compact topological space in which every element is a limit point of the set.

To construct the *Cantor set* as a subset of $[0, 1]$, let $I_0 = [0, 1] \subset \mathbf{R}^1$, $I_1 = [0, \frac{1}{3}] \cup [\frac{2}{3}, 1]$ and $I_2 = [0, \frac{1}{9}] \cup [\frac{2}{9}, \frac{1}{3}] \cup [\frac{2}{3}, \frac{7}{9}] \cup [\frac{8}{9}, 1]$. In general, define I_n to be the union of closed intervals obtained by removing the open "middle thirds"

from each of the closed intervals comprising I_{n-1}. The Cantor set is defined as $C = \cap_{n=1}^{\infty} I_n$.

The Cantor set has length 0, which can be verified by summing the lengths of the intervals removed to obtain a sum of 1. It is a closed set where each point is an accumulation point. On the other hand, it can be shown that the Cantor set can be placed in one-to-one correspondence with the points of the interval $[0, 1]$.

(2) Any topological space homeomorphic to the standard Cantor set in $\mathbf{R}^1$.

Cantor's Theorem   If S is any set, there is no surjection from S onto the power set $\mathcal{P}(S)$.

Cartan formula   A formula expressing the relationship between values of an operation on a product of terms and products of operations applied to individual terms. For the mod 2 Steenrod algebra, the *Cartan formula* is given by

$$Sq^i(xy) = \sum_j (Sq^j x)(Sq^{i-j} y).$$

The sum is finite since $Sq^j x = 0$ when j is greater than the degree of the cohomology class x. A differential in a spectral sequence is another example where there is a Cartan formula (if there is a product on the spectral sequence).

Cartesian product   For any two sets X and Y, the set, denoted $X \times Y$, of all ordered pairs (x, y) with $x \in X$, $y \in Y$.

Cartesian space   The standard coordinate space $\mathbf{R}^n$, where points are given by n real-valued coordinates for some n. Distance between two points $x = (x_1, \ldots, x_n)$ and $y = (y_1, \ldots, y_n)$ is determined by the Pythagorean identity:

$$d(x, y) = \sqrt{\sum_{i=1}^{n} (x_i - y_i)^2}.$$

Cartesian space is a model of Euclidean geometry.

catastrophe theory   The study of quantities which may change suddenly (discontinuously) even while the quantities that determine them change smoothly.

Example: When forces on an object grow to the point of overcoming the opposing force due to friction, the object will move suddenly.

categorical theory   A consistent theory T in a language L is *categorical* if all models of T are isomorphic. Because of the Löwenheim-Skolem Theorem, no theory with an infinite model can be categorical in this sense, since models of different cardinalities cannot be isomorphic.

More generally, a consistent theory T is κ-categorical for a cardinal κ if any two models of T of size κ are isomorphic.

category   A *category* **X** consists of a class of objects, Obj(**X**), pairwise disjoint sets of functions (morphisms), $\text{Hom}_X(A, B)$, for every ordered pair of objects $A, B \in \text{Obj}(\mathbf{X})$, and compositions

$$\text{Hom}_X(A, B) \times \text{Hom}_X(B, C) \to \text{Hom}_X(A, C),$$

denoted $(f, g) \mapsto gf$ satisfying the following properties:

(i.) for each $A \in \text{Obj}(\mathbf{X})$ there is an identity morphism $1_A \in \text{Hom}_C(A, A)$ such that $f1_A = f$ for all $f \in \text{Hom}_X(A, B)$ and $1_A g = g$ for all $g \in \text{Hom}_X(C, A)$;

(ii) associativity of composition for morphisms holds whenever possible: if $f \in \text{Hom}_X(A, B)$, $g \in \text{Hom}_X(B, C)$, $h \in \text{Hom}_X(C, D)$, then $h(gf) = (hg)f$.

category of groups   The class of all groups $G, H, \ldots$, with each $\text{Hom}(G, H)$ equal to the set of all group homomorphisms $f : G \to H$, under the usual composition. Denoted **Grp**. *See* category.

category of linear spaces   The class of all vector spaces $V, W, \ldots$, with each $\text{Hom}(V, W)$ equal to the set of all linear transformations $f : V \to W$, under the usual composition. Denoted **Lin**. *See* category.

category of manifolds   The class of all differentiable manifolds $M, N, \ldots$, with each $\text{Hom}(M, N)$ equal to the set of all differentiable functions $f : M \to N$, under the usual composition. Denoted **Man**. *See* category.

category of rings   The class of all rings $R, S, \ldots$, with each $\text{Hom}(R, S)$ equal to the set of all ring homomorphisms $f : R \to S$, under the usual composition. Denoted **Ring**. *See* category.

category of sets   The class of all sets $X, Y, \ldots$, with $\text{Hom}(X, Y)$ equal to the set of all functions $f : X \to Y$, under the usual composition. Denoted **Set**. *See* category.

category of topological spaces   The class of all topological spaces $X, Y, \ldots$, with each $\text{Hom}(X, Y)$ equal to the set of all continuous functions $f : X \to Y$, under the usual composition. Denoted **Top**. *See* category.

Cauchy sequence   An infinite sequence $\{x_n\}$ of points in a metric space M, with distance function d, such that, given any positive number ϵ, there is an integer N such that for any pair of integers m, n greater than N the distance $d(x_m, x_n)$ is always less than ϵ. Any convergent sequence is automatically a *Cauchy sequence*.

Cavalieri's Theorem   The theorem or principle that if two solids have equal area cross-sections, then they have equal volumes, was published by Bonaventura Cavalieri in 1635. As a consequence of this theorem, the volume of a cylinder, even if it is oblique, is determined only by the height of the cylinder and the area of its base.

cell   A set whose interior is homeomorphic to the n-dimensional unit disk $\{x \in \mathbf{R}^n : \|x\| < 1\}$ and whose boundary is divided into finitely many lower-dimensional cells, called *faces* of the original *cell*. The number n is the dimension of the cell and the cell itself is called an n-cell. Cells are the building blocks of a complex.

central symmetry   The property of a geometric figure F, such that F contains a point c (the *center of* F) so that, for every point p_1 on F, there is another point p_2 on F such that c bisects the line segment $\overline{p_1 p_2}$.

centroid   The point of intersection of the three medians of a triangle.

chain   A formal finite linear combination of simplices in a simplicial complex K with integer coefficients, or more generally with coefficients in some ring. The term is also used in more general settings to denote an element of a *chain complex*.

chain complex   Let R be a ring (for example, the integers). A *chain complex* of R-modules consists of a family of R-modules C_n, where n ranges over the integers (or sometimes the non-negative integers), together with homomorphisms $\partial_n : C_n \longrightarrow C_{n-1}$ satisfying the condition: $\partial_{n-1} \circ \partial_n(x) = 0$ for every x in C_n.

chain equivalent complexes   Let $C = \{C_n\}$ and $C' = \{C'_n\}$ be chain complexes with boundary maps ∂ and ∂', respectively. (*See* chain complex.) A chain mapping $f : C \longrightarrow C'$ is a *chain equivalence* if there is a chain mapping $g : C' \longrightarrow C$ and chain homotopies from $g \circ f$ to the identity mapping of C and from $f \circ g$ to the identity mapping of C'. In this case we say that C and C' are chain equivalent. A chain equivalence induces an isomorphism between the homology of C and the homology of C'. For example, if $\phi : X \longrightarrow Y$ is a homotopy equivalence of topological spaces, then ϕ induces a chain equivalence of the singular chain complexes of X and Y.

chain group   Let K be a simplicial complex. Then the nth *chain group* $C_n(K)$ is the free Abelian group constructed by taking all finite linear combinations with integer coefficients of n-dimensional simplices of K. Similarly, if X is a topological space, the nth singular chain group is the free Abelian group constructed by taking all finite linear combinations of singular simplices, which are continuous functions from the standard n-dimensional simplex to X.

chain homotopy   Let $C = \{C_n\}$ and $C' = \{C'_n\}$ be chain complexes with boundary maps ∂_n and ∂'_n, respectively. Let f and g be chain mappings from C to C'. *See* chain complex, chain mapping. Then a *chain homotopy* T from f to g is a collection of homomorphisms $T_n : C_n \longrightarrow C'_{n+1}$ such that $\partial_{n+1} \circ T_n + T_{n-1} \circ \partial_n = f_n - g_n$. For example, a homotopy between two maps from a topological space X to a topologi-cal space Y induces a chain homotopy between the induced chain maps from the singular chain complex of X to the singular chain complex of Y.

chain mapping   Let $C = \{C_n\}$ and $C' = \{C'_n\}$ be chain complexes with boundary maps $\partial_n : C_n \longrightarrow C_{n-1}$ and $\partial'_n : C'_n \longrightarrow C'_{n-1}$, respectively. *See* chain complex. A *chain mapping* $f : C \longrightarrow C'$ is a family of homomor-phisms $f_n : C_n \longrightarrow C'_n$ satisfying $\partial'_n \circ f_n = f_{n-1} \circ \partial_n$. For example, when $\phi : X \longrightarrow Y$ is continuous, the induced map from the singular chain complex of X to the singular chain com-plex of Y is a chain map.

characteristic class   Let $E \longrightarrow B$ be a vector bundle. A *characteristic class* assigns a class ξ in the cohomology $H^*(B)$ of B to each vector bundle over B so that the assignment is "pre-dictable" or natural with respect to maps of vec-tor bundles. That is, if the maps $f : E \longrightarrow E'$ and $g : B \longrightarrow B'$ form a map of vector bundles so that $E \longrightarrow B$ is equivalent to the pullback $g^*(E') \longrightarrow B$, then the class assigned to $E \longrightarrow B$ is the image of the class assigned to $E' \longrightarrow B'$ under the map $g^* : H^*(B') \longrightarrow H^*(B)$.

When the cohomology of the base space can be considered as a set of numbers, the charac-teristic class is sometimes called a *characteristic number*.

Example: Stiefel-Whitney classes of a man-ifold are characteristic classes in mod 2 coho-mology.

characteristic function   The *characteristic function* χ_A of a set A of natural numbers is the function that indicates membership in that set; i.e., for all natural numbers n,

$$\chi_A(n) = \begin{cases} 1 & \text{if } n \in A \\ 0 & \text{if } n \notin A. \end{cases}$$

More generally, if A is a fixed universal set and $B \subseteq A$, then for all $x \in A$,

$$\chi_B(x) = \begin{cases} 1 & \text{if } x \in B \\ 0 & \text{if } x \notin B. \end{cases}$$

characteristic number   *See* characteristic class.

choice function   Suppose that $\{X_\alpha\}_{\alpha\in\Gamma}$ is a family of non-empty sets. A *choice function* is a function $f : \{X_\alpha\}_{\alpha\in\Gamma} \to \bigcup_{\alpha\in\Gamma} X_\alpha$ such that $f(X_\alpha) \in X_\alpha$ for all $\alpha \in \Gamma$. *See also* Axiom of Choice.

choice set   Suppose that $\{X_\alpha\}_{\alpha\in\Gamma}$ is a family of pairwise disjoint, non-empty sets. A *choice set* is a set Y, which consists of exactly one element from each set in the family. *See also* Axiom of Choice.

chord   A line segment with endpoints on a curve (usually a circle).

Christoffel symbols   The coefficients in local coordinates for a connection on a manifold. If $(u^1, \ldots, u^n)$ is a local coordinate system in a manifold M and ∇ is a covariant derivative operator, then the derivatives of the coordinate fields $\frac{\partial}{\partial u^j}$ can be written as linear combinations of the coordinate fields:

$$\nabla_{\frac{\partial}{\partial u^i}} \frac{\partial}{\partial u^j} = \sum_{k=1}^{n} \Gamma_{ij}^{k} \frac{\partial}{\partial u^k}.$$

The functions $\Gamma_{ij}^{k}(u^1, \ldots, u^n)$ are the *Christoffel symbols*. For the standard connection on Euclidean space $\mathbf{R}^n$ the Christoffel symbols are identically zero in rectilinear coordinates, but in general coordinate systems they do not vanish even in $\mathbf{R}^n$.

Church-Turing Thesis   If a partial function φ on the natural numbers is computable by an algorithm in the intuitive sense, then φ is computable, in the formal, mathematical sense. (A function φ on the natural numbers is *partial* if its domain is some subset of the natural numbers.) *See* computable.

This statement of the *Church-Turing Thesis* is a modern day rephrasing of independent statements by Alonzo Church and Alan Turing. Church's Thesis, published by Church in 1936, states that the intuitively computable partial functions are exactly the general recursive functions, where the notion of general recursive function is a formalization of computable defined by Gödel. Turing's Thesis, published by Turing in 1936, states that the intuitively computable partial functions are exactly the partial functions which are Turing computable.

The Church-Turing Thesis is a statement that cannot be proved; rather it must be accepted or rejected. The Church-Turing Thesis is, in general, accepted by mathematicians; evidence in favor of accepting the thesis is that all known methods of formalizing the notion of computability (*see* computable) have resulted in the same class of functions; i.e., a partial function φ is partial recursive if and only if it is Turing computable, etc.

The most important use of the Church-Turing Thesis is to define formally the notion of non-computability. To show the lack of *any* algorithm to compute a function, it suffices by the thesis to show that the function is not partial recursive (or Turing computable, etc.). The converse of the Church-Turing Thesis is clearly true.

circle   The curve consisting of all points in a plane which are a fixed distance (the radius of the circle) from a fixed point (the center of the circle) in the plane.

circle of curvature   For a plane curve, a *circle of curvature* is the circle defined at a point on the curve that is both tangent to the curve and has the same curvature as the curve at that point. For a space curve, the osculating circle is the circle of curvature.

circle on sphere   The intersection of the surface of the sphere with a plane.

circular arc   A segment of a circle.

circular cone   A cone whose base is a circle.

circular cylinder   A cylinder whose bases are circles.

circular helix   A curve lying on the surface of a circular cylinder that cuts the surface at a constant angle. It is parameterized by the equations $x = a \sin t$, $y = a \cos t$, and $z = bt$, where a and b are real constants.

circumcenter of triangle   The center of a circle circumscribed about a given triangle. The circumcenter coincides with the point common to the three perpendicular bisectors of the triangle. *See* circumscribe.

circumference of a circle The perimeter, or length, of a circle.

circumference of a sphere The circumference of a great circle of the sphere. *See* circumference of a circle, great circle.

circumscribe Generally a plane (or solid) figure *F circumscribes* a polygon (or polyhedron) *P* if the region bounded by *F* contains the region bounded by *P* and if every vertex of *P* is incident with *F*. In such a case *P* is said to be *inscribed in F*. *See* circumscribed circle, for example. In specific circumstances, figures other than polygons and polyhedra may also be circumscribed.

circumscribed circle A circle containing the interior of a polygon in its interior, in such a way that every vertex of the polygon is on the circle; i.e., the polygon is inscribed in the circle.

circumscribed cone A cone that circumscribes a pyramid in such a way that the base of the cone circumscribes the base of the pyramid and the vertex of the cone coincides with the vertex of the pyramid; i.e., the pyramid is inscribed in the cone. *See* circumscribe.

circumscribed cylinder A cylinder that circumscribes a prism in such a way that both bases of the cylinder circumscribe a base of the prism; i.e., the prism is inscribed in the cylinder. *See* circumscribe.

circumscribed polygon A polygon that contains the region bounded by a closed curve (usually a circle) in the region it bounds, in such a way that every side of the polygon is tangent to the closed curve; i.e., the closed curve is inscribed in the polygon.

circumscribed polyhedron A polyhedron that bounds a volume containing the volume bounded by a closed surface (usually a sphere) in such a way that every face of the polyhedron is tangent to the closed surface; i.e., the closed surface is inscribed in the polyhedron. *See* circumscribe.

circumscribed prism A prism that contains the interior of a cylinder in its interior, in such a way that both bases of the prism circumscribe a base of the cylinder (and so each lateral face of the prism is tangent to the cylindrical surface); i.e., the cylinder is inscribed in the prism. *See* circumscribe.

circumscribed pyramid A pyramid that contains, in its interior, the interior of a cone, in such a way that the base of the pyramid circumscribes the base of the cone and the vertex of the pyramid coincides with the vertex of the cone; i.e., the cone is inscribed in the pyramid. *See* circumscribe.

circumscribed sphere A sphere that contains, in its interior, the region bounded by a polyhedron, in such a way that every vertex of the polyhedron is on the sphere; i.e., the polyhedron is inscribed in the sphere. *See* circumscribe.

class The collection of all objects that satisfy a given property. Every set is a *class,* but the converse is not true. A class that is not a set is called a *proper class*; such a class is much "larger" than a set because it cannot be assigned a cardinality. *See* Bernays-Gödel set theory.

classifying space The *classifying space* of a topological group G is a space BG with the property that the set of equivalence classes of vector bundles $p : E \longrightarrow B$ with G-action is in bijective correspondence with the set $[B, BG]$ of homotopy classes of maps from the space B to BG.

The space BG is unique up to homotopy, that is, any two spaces satisfying the above property for a fixed group G are homotopy equivalent.

For $G = \mathbf{Z}/2$, $B\mathbf{Z}/2$ is an infinite projective space $\mathbf{RP}^{\infty}$, the union of all projective spaces $\mathbf{RP}^n$. Since $O(1) = \mathbf{Z}/2$, all line bundles over a space X are classified up to bundle homotopy equivalence by homotopy classes of maps from X into $\mathbf{RP}^{\infty}$.

closed and unbounded If κ is a non-zero limit ordinal (in practice κ is an uncountable cardinal), and $C \subseteq \kappa$, C is *closed and unbounded* if it satisfies (i.) for every sequence

$\alpha_0 < \alpha_1 < \cdots < \alpha_\beta \ldots$ of elements of C (where $\beta < \gamma$, for some $\gamma < \kappa$), the supremum of the sequence, $\bigcup_{\beta < \gamma} \alpha_\beta$, is in C, and (ii.) for every $\alpha < \kappa$, there exists $\beta \in C$ such that $\beta > \alpha$. A closed and unbounded subset of κ is often called a *club subset* of κ.

closed convex curve   A curve C in the plane which is a closed curve and is the boundary of a convex figure A. That is, the line segment joining any two points in C lies entirely within A. Equivalently, if A is a closed bounded convex figure in the plane, then its boundary C is a *closed convex curve*.

closed convex surface   The boundary S of a closed convex body in three-dimensional Euclidean space. S is topologically equivalent to a sphere and the line segment joining any two points in S lies in the bounded region bounded by S.

closed formula   A well-formed formula φ of a first-order language such that φ has no free variables.

closed half line   A set in $\mathbf{R}$ of the form $[a, \infty)$ or $(-\infty, a]$ for some $a \in \mathbf{R}$.

closed half plane   A subset of $\mathbf{R}^2$ consisting of a straight line L and exactly one of the two half planes which L determines. Thus, any *closed half plane* is either of the form $\{(x, y) : ax + by \geq c\}$ or $\{(x, y) : ax + by \leq c\}$. The sets $x \geq c$ and $x \leq c$ are vertical closed half planes; $y \geq c$ and $y \leq c$ are horizontal half planes.

closed map   A function $f : X \to Y$ between topological spaces X and Y such that, for any closed set $C \subseteq X$, the image set $f(C)$ is closed in Y.

closed set   (1) A subset A of a topological space, such that the complement of A is open. *See* open set. For example, the sets $[a, b]$ and $\{a\}$ are closed in the usual topology of the real line.

(2) A *closed set* of ordinals is one that is closed in the order topology. That is, $C \subseteq \kappa$ is closed if, for any limit ordinal $\lambda < \kappa$, if $C \cap \lambda$ is unbounded in λ, then $\lambda \in C$. Equivalently, if

$\{\beta_\alpha : \alpha < \lambda\} \subseteq C$ is an increasing sequence of length $\lambda < \kappa$, then

$$\beta = \lim_{\alpha \to \lambda} \beta_\alpha \in C .$$

For example, the set of all limit ordinals less than κ is closed in κ. *See also* unbounded set, stationary set.

closed surface   A compact Hausdorff topological space with the property that each point has a neighborhood topologically equivalent to the plane. Thus, a *closed surface* is a compact 2-dimensional manifold without boundary. The ellipsoids given by $\frac{x^2}{a^2} + \frac{y^2}{b^2} + \frac{z^2}{c^2} - 1 = 0$ are simple examples of closed surfaces. More generally, if $f(x, y, z)$ is a differentiable function, then the set of points S satisfying $f(x, y, z) = 0$ is a closed surface provided that S is bounded and the gradient of f does not vanish at any point in S.

closure of a set   The *closure* of a subset A of a topological space X is the smallest closed set $\bar{A} \subseteq X$ which contains A. In other words, $\bar{A}$ is the intersection of all closed sets in X that contain A. Equivalently, $\bar{A} = A \cup A'$, where A' is the derived set of A. For example, the closure of the rationals in the usual topology is the whole real line.

cluster point   *See* accumulation point.

cobordism   A *cobordism* between two n-dimensional manifolds is an $(n+1)$-dimensional manifold whose boundary is the disjoint union of the two lower dimensional manifolds. A cobordism between two manifolds with a certain structure must also have that structure. For example, if the manifolds are real oriented manifolds, then the cobordism must also be a real oriented manifold.

Example: The cylinder provides a cobordism between the circle and itself. Any manifold with boundary provides a cobordism between the boundary manifold and the empty set, which is considered an n-manifold for all n.

cobordism class   For a manifold M, the class of all manifolds cobordant M, that is, all manifolds N for which there exists a manifold W

whose boundary is the disjoint union of M and N.

cobordism group   The cobordism classes of n-dimensional manifolds (possibly with additional structure) form an Abelian group; the product is given by disjoint union. The identity element is the class given by the empty set. The inverse of the cobordism class of a manifold M is given by reversing the orientation of M; the manifold $M \times [0, 1]$ is a cobordism between M and M with the reverse orientation. (*See* cobordism class.) When studying cobordism classes of unoriented manifolds, each manifold is its own inverse; thus, all such cobordism classes are 2-torsion.

Some results in geometry show that cobordant manifolds may have a common geometric or topological property, for example, two spincobordant manifolds either both admit a positive scalar curvature metric, or neither manifold can have such a metric.

Codazzi-Mainardi equations   A system of partial differential equations arising in the theory of surfaces. If M is a surface in $\mathbf{R}^3$ with local coordinates (u^1, u^2), its geometric invariants can be described by its first fundamental form $g_{ij}(u^1, u^2)$ and second fundamental form $L_{ij}(u^1, u^2)$. The Christoffel symbols Γ^k_{ij} are determined by the first fundamental form. (*See* Christoffel symbols.) In order for functions g_{ij} and L_{ij}, $i, j = 1, 2$ to be the first and second fundamental forms of a surface, certain integrability conditions (arising from equality of mixed partial derivatives) must be satisfied. One set of conditions, the *Codazzi-Mainardi equations,* is given in terms of the Christoffel symbols by:

$$\frac{\partial L_{ik}}{\partial u^j} - \frac{\partial L_{ij}}{\partial u^k} + \Gamma^l_{ik} L_{lj} - \Gamma^l_{ij} L_{lk} = 0 \,.$$

codimension   A nonnegative integer associated with a subspace W of a space V. Whenever the space has a dimension (e.g., a topological or a vector space) denoted by $\dim V$, the *codimension* of W is the defect $\dim V - \dim W$. For example, a curve in a surface has codimension 1 (topology) and a line in space has codimension 2 (a line through the origin is a vector subspace $\mathbf{R}$ of $\mathbf{R}^3$).

cofinal   Let α, β be limit ordinals. An increasing sequence $\langle \alpha_\tau : \tau < \beta \rangle$ is *cofinal* in α if $\lim_{\tau \to \beta} \alpha_\tau = \alpha$. *See* limit ordinal.

cofinality   Let α be an infinite limit ordinal. The *cofinality* of α is the least ordinal β such that there exists a sequence $\langle \alpha_\tau : \tau < \beta \rangle$ which is cofinal in α. *See* cofinal.

cofinite subset   A subset A of an infinite set S, such that $S \backslash A$ is finite. Thus, the set of all integers with absolute value at least 13 is a *cofinite subset* of $\mathbf{Z}$.

coimage   Let $\mathcal{C}$ be an additive category and $f \in \mathrm{Hom}_{\mathcal{C}}(X, Y)$ a morphism. If $i \in \mathrm{Hom}_{\mathcal{C}}(X', X)$ is a morphism such that $fi = 0$, then a *coimage* of f is a morphism $g \in \mathrm{Hom}_{\mathcal{C}}(X, Y')$ such that $gi = 0$. *See* additive category.

coinfinite subset   A subset A if an infinite set S such that $S \backslash A$ is infinite. Thus, the set of all even integers is a *coinfinite subset* of $\mathbf{Z}$.

collapse   A *collapse of a complex K* is a finite sequence of elementary combinatorial operations which preserves the homotopy type of the underlying space.

For example, let K be a simplicial complex of dimension n of the form $K = L \cup \sigma \cup \tau$, where L is a subcomplex of K, σ is an open n-simplex of K, and τ is a free face of σ. That is, τ is an $n - 1$ dimensional face of σ and is not the face of any other n-dimensional simplex.

The operation of replacing the complex $L \cup \sigma \cup \tau$ with the subcomplex L is called an elementary collapse of K and is denoted $K \searrow L$. A collapse is a finite sequence of elementary collapses $K \searrow L_1 \cdots \searrow L_m$.

When K is a CW complex, ball pairs of the form (B^n, B^{n-1}) are used in place of the pair (σ, τ).

collection   *See* set.

collinear   Points that lie on the same line or on planes that share a common line.

comb space   A topological subspace of the plane $\mathbf{R}^2$ which resembles a comb with infinitely many teeth converging to one end. For example, the subset of the unit square $[0, 1]^2$ given by

$$(\{0\} \times [0, 1]) \cup$$
$$(\{\frac{1}{k} : k \geq 1\} \times [0, 1]) \cup ([0, 1] \times \{0\})$$

is a *comb space*. The subspace obtained from this set by deleting the line segment $\{0\} \times (0, 1)$ is an example of a connected set that is not path connected.

common tangent   A line that is tangent to two circles.

commutative diagram   A diagram

$$
\begin{array}{ccc}
A & \xrightarrow{f_1} & B \\
{\scriptstyle g_1}\downarrow & & \downarrow{\scriptstyle g_2} \\
C & \xrightarrow{f_2} & D
\end{array}
$$

in which the two compositions $g_2 f_1$ and $f_2 g_1$ are equal. Commutative triangles can be considered a special case if one of the functions is the identity. Larger diagrams composed of squares and triangles commute if each square and triangle inside the diagram commutes. *See* diagram.

compact   **(1)** The property of a topological space X that every cover of X by open sets (every collection $\{X_\alpha\}$ of open sets with $X \subset \cup X_\alpha$) contains a finite subcover (a finite collection $X_{\alpha_1}, \ldots, X_{\alpha_n}$ with $X \subset \cup X_{\alpha_i}$).
    (2) A compact topological space.

compact complex manifold   A complex manifold which is compact in the complex topology. A common example is a Riemann surface (1-dimensional complex manifold): the (Riemann) sphere is compact, unlike the sphere with a point removed. The sphere with an open disk removed is also compact in the complex topology, but strictly speaking it is not a complex manifold (some points do not lie in an open disk): it is known as a manifold with boundary. *See* complex manifold.

compactification   A *compactification* of a topological space X is a pair, (Y, f), where Y is a compact Hausdorff space and f is a homeomorphism from X onto a dense subset of Y. A

necessary and sufficient condition for a space to have a compactification is that it be completely regular. *See also* one-point compactification, Stone-Čech compactification.

compact leaf   A concept arising in the theory of foliations. A foliated manifold is an n-dimensional manifold M, partitioned into a family of disjoint, path-connected subsets L_α such that there is a covering of M by open sets U_i and homeomorphisms $h_i : U_i \longrightarrow \mathbf{R}^n$ taking each component of $L_\alpha \cap U_i$ onto a parallel translate of the subspace $\mathbf{R}^k$. Each L_α is called a leaf, and it is a *compact leaf* if it is compact as a subspace.

compact-open topology   The topology on the space of continuous functions from a topological space X to a topological space Y, generated by taking as a subbasis all sets of the form $\{f : f(C) \subseteq U\}$, where $C \subseteq X$ is compact and $U \subseteq Y$ is open. If Y is a metric space, this topology is the same as that given by uniform convergence on compact sets.

comparability of cardinal numbers   The proposition that, for any two cardinals α, β, either $\alpha \leq \beta$ or $\beta \leq \alpha$.

compass   An instrument for constructing points at a certain distance from a fixed point and for measuring distance between points.

compatible (elements of a partial ordering)   Two elements p and q of a partial order $(\mathcal{P}, \leq)$ such that there is an $r \in \mathcal{P}$ with $r \leq p$ and $r \leq q$. Otherwise p and q are incompatible.
    In the special case of a Boolean algebra, p and q are compatible if and only if $p \wedge q \neq 0$. In a tree, however, p and q are compatible if and only if they are comparable: $p \leq q$ or $q \leq p$.

complementary angles   Two angles are *complementary* if their sum is a right angle.

complement of a set   If X is a set contained in a universal set U, the *complement of* X, denoted X', is the set of all elements in U that do not belong to X. More precisely, $X' = \{u \in U : u \notin X\}$.

completely additive function   An arithmetic function f having the property that $f(mn) = f(m) + f(n)$ for all positive integers m and n. (*See* arithmetic function.) For example, the function $f(n) = \log n$ is completely additive. The values of a completely additive function depend only on its values at primes, since $f(p^i) = i \cdot f(p)$. *See also* additive function.

completely multiplicative function   An arithmetic function f having the property that $f(mn) = f(m) \cdot f(n)$ for all positive integers m and n. (*See* arithmetic function.) For example, λ, Liouville's function, is completely multiplicative. The values of a completely multiplicative function depend only on its values at primes, since $f(p^i) = (f(p))^i$. *See also* multiplicative function, strongly multiplicative function.

completely normal topological space   A topological space X such that every subspace of X is normal. In particular, X itself must be normal, and since normality is not generally preserved in subspaces, complete normality is stronger than normality. Complete normality is equivalent to requiring that, for all subsets A and B of X, if $\bar{A} \cap B = A \cap \bar{B} = \emptyset$, then there are disjoint open sets U and V with $A \subseteq U$ and $B \subseteq V$.

completely regular topological space   A topological space X such that points are closed and points and closed sets can be separated by continuous functions. That is, for each $x \in X$, the singleton $\{x\}$ is closed, and for all closed $C \subseteq X$ with $x \notin C$, there is a continuous $f : X \to [0, 1]$ such that $f(x) = 0$ and $f(c) = 1$ for all $c \in C$.

complete metric space   A topological space X with metric d such that any Cauchy sequence in X converges. That is, if $\{x_n : n \in \mathbf{N}\} \subseteq X$ is such that for any $\epsilon > 0$ there is an N with $d(x_n, x_m) < \epsilon$ for any $n, m \geq N$, then there is an $x \in X$ with $x_n \to x$. For example, each Euclidean n-space $\mathbf{R}^n$ is a *complete metric space*.

complete set of logical connectives   A set C of logical connectives such that, given any well-formed propositional (sentential) formula φ, whose logical connectives are from among the usual set $\{\neg, \wedge, \vee, \to, \leftrightarrow\}$ of logical connectives, there is a well-formed propositional formula ψ, whose logical connectives are from C, such that φ and ψ are logically equivalent.

Examples of *complete sets of logical connectives* include $\{\neg, \wedge, \vee\}$, $\{\neg, \wedge\}$, $\{\neg, \vee\}$, and $\{\neg, \to\}$. The set $\{\wedge, \to\}$ is not a complete set of logical connectives.

complete theory   Let $\mathcal{L}$ be a first order language and let T be a (closed) theory of $\mathcal{L}$. The theory T is *complete* if, for all sentences σ, either $\sigma \in T$ or $(\neg\sigma) \in T$.

If T is simply a collection of sentences, then T is complete if for all sentences σ, either σ is a logical consequence of T or $(\neg\sigma)$ is a logical consequence of T. Equivalently, T is complete if, for all sentences σ, either σ is provable from T or $(\neg\sigma)$ is provable from T.

Let $\mathcal{A}$ be a structure for $\mathcal{L}$. The theory of $\mathcal{A}$ (denoted $Th(\mathcal{A})$), the set of all sentences of $\mathcal{L}$ which are true in $\mathcal{A}$, is a *complete theory*.

complex   A collection of cells with the properties: (i.) if C is a cell in the complex, then every face of C is in the complex; and (ii.) every two cells in the complex have disjoint interiors.

complex analytic fiber bundle   A fiber bundle $f : F \to X$ where F and X are complex manifolds and f is an analytic map. *See* fiber bundle.

complex analytic structure   On a real differential manifold M an integrable complex structure on the tangent bundle TM; namely, the data of an invertible linear map $J_p : T_pM \to T_pM$ on each tangent space at $p \in M$, such that $J_p^2 = -$ Identity, which varies smoothly with p and is integrable, i.e., admits an atlas with constant transition functions. Without the integrability condition, the data define an "almost-complex structure" on M.

complex conjugate bundle   For a complex vector bundle $f : V \to M$, the *conjugate bundle* $\overline{V}$ is defined by taking the complex conjugate $\overline{f}_\alpha$ of each local map $f_\alpha : \mathbf{C}^n \times U_\alpha \cong V|_{U_\alpha} = f^{-1}(U_\alpha)$ that defines the bundle restricted to U_α, for a suitable covering U_α of M.

complex dimension (1) For a vector space X, the dimension of X, considered as a vector space over the field $\mathbf{C}$ of complex numbers, as opposed to the *real dimension*, which is the dimension of X as a vector space over the real numbers $\mathbf{R}$.

(2) (For a complex manifold M). The complex dimension of the tangent space $T_p M$ at each point p.

(3) The dimension of a complex; i.e., the highest of the dimensions of the cells that form the complex.

complex line bundle A complex vector bundle whose fibers have dimension 1. *See* complex vector bundle.

complex manifold A set of points M which can be covered with a family of subsets $\{U_\alpha\}_{\alpha \in A}$ i.e., $M = \bigcup_{\alpha \in A} U_\alpha$, each of which is isomorphic to an open ball in complex n-space: $\{(z_1, \ldots, z_n) \in \mathbf{C}^n : |z_1|^2 + \ldots + |z_n|^2 = 1\}$, for a fixed non-negative integer n.

complex of lines In projective geometry, a *line complex* is a subvariety of the Grassmannian $\mathrm{Gr}(2, 4)$ of all lines in (complex) projective 3-space $\mathbf{CP}^3$, which is the set of 2-dimensional subspaces of a 4-dimensional complex vector space. $\mathrm{Gr}(2, 4)$ is a quadric hypersurface in $\mathbf{CP}^5$, thus an example of line complex is a "linear line complex", the intersection of $\mathrm{Gr}(2, 4)$ with a hyperplane, e.g., all the lines in $\mathbf{P}^3$ that meet a given plane.

complex plane The topological space, denoted $\mathbf{C}$ or $\mathbb{C}$, consisting of the set of complex numbers, i.e., numbers of the form $a + bi$, where a and b are real numbers and $i^2 = -1$. $\mathbf{C}$ is usually visualized as the set of pairs (a, b) and hence the terminology *plane*.

The term *extended* complex plane refers to $\mathbf{C}$, together with a point at infinity and neighborhoods of the form $\{z : \|z\| > r\}$ for real numbers r.

complex sphere (1) A sphere $\{z : |z - z_0| = r\}$, in the complex plane.

(2) A unit sphere whose points are identified with points in the complex plane by a stereographic projection, with the "north pole" identified with the point ∞. Such a sphere, therefore,

represents the extended complex plane. *See* complex plane.

complex torus The n-dimensional compact complex analytic manifold $\mathbf{C}^n / \Lambda$, where n is a positive integer and Λ a complete lattice in $\mathbf{C}^n$. In dimension 1, the complex tori $\mathbf{C}/(\mathbf{Z}\omega_1 + \mathbf{Z}\omega_2)$, where ω_1 and ω_2 are complex numbers independent over $\mathbf{R}$, are all algebraic varieties, also called elliptic curves.

complex vector bundle A *complex vector bundle* (of dimension n) on a differentiable manifold M is a manifold E, given by a family of complex vector spaces $\{E_p\}_{p \in M}$, with a trivialization over an open covering $\{U_\alpha\}_{\alpha \in A}$ of M, namely diffeomorphisms $\phi_\alpha : \mathbf{C}^n \times U_\alpha \to \{E_p\}_{p \in U_\alpha}$. If M and E are complex analytic manifolds and a trivialization exists with ϕ_α biholomorphic maps, the bundle is said to be complex analytic.

composite *See* composite number.

composite number An integer, other than -1, 0, and 1, that is not a prime number. That is, a nonzero integer is composite if it has more than two positive divisors. For example, 6 is composite since the positive divisors of 6 are 1, 2, 3, and 6. Just as prime numbers are usually assumed to be positive integers, a *composite number* is usually assumed to be positive as well.

composition of functions Suppose that $f : X \to Y$ and $g : Y \to Z$ are functions. The composition $gf : X \to Z$ is the function consisting of all ordered pairs (x, z) such that there exists an element $y \in Y$ with $(x, y) \in f$ and $(y, z) \in g$. *See* function.

computable Let $\mathbf{N}$ be the set of natural numbers. Intuitively, a function $f : \mathbf{N} \to \mathbf{N}$ is *computable* if there is an algorithm, or effective procedure, which, given $n \in \mathbf{N}$ as input, produces $f(n)$ as output in finitely many steps. There are no limitations on the amount of time or "memory" (i.e., "scratch paper") necessary to compute $f(n)$, except that they be finite. If $f : \mathbf{N}^k \to \mathbf{N}$, then f is computable is defined analogously.

A function φ on $\mathbf{N}$ is *partial* if its domain is some subset of $\mathbf{N}$; i.e., φ may not be defined on all inputs. A partial function φ on $\mathbf{N}$ is intuitively computable if there is an algorithm, or effective procedure, which given $n \in \mathbf{N}$ as input, produces $\varphi(n)$ as output in finitely many steps if $n \in \mathrm{dom}(\varphi)$, and runs forever otherwise.

For example, the function $f(n, m) = n + m$ is intuitively computable, as is the function f which, on input $n \in \mathbf{N}$, produces as output the nth prime number. The function φ which, on input $n \in \mathbf{N}$, produces the output 1 if there exists a consecutive run of exactly n 5s in the decimal expansion of π, and is undefined otherwise, is an intuitively computable partial function.

The notion of computability has a formal mathematical definition; in order to say that a function is *not* computable, one must have a formal mathematical definition. There have been several formalizations of the intuitive notion of computability, all of which generate the same class of functions. Given here is the formalization of Turing computable. A second formalization is given in the definition of a partial recursive function. *See* partial recursive function. Other formalizations include that of register machine computability (Shepherdson–Sturgis, 1963), general recursive functions (Gödel, 1934), and λ-definable functions (Church, 1930). It has been proved that, for any partial function φ, φ is Turing computable if and only if φ is partial recursive, if and only if φ is register machine computable, etc. *See also* Church-Turing Thesis. Thus, the term *computable* can (mathematically) mean computable in any such formalization.

A set A of natural numbers is computable if its characteristic function is computable; i.e., the function

$$\chi_A(n) = \begin{cases} 1 & \text{if } n \in A \\ 0 & \text{if } n \notin A \end{cases}$$

is recursive.

A partial function φ on $\mathbf{N}$ is Turing computable if there is some Turing machine that computes it. The notion of Turing machine was formalized by Alan Turing in his 1936 *Proceedings of the London Mathematical Society* paper.

A Turing machine consists of a bi-infinite tape, which is divided into cells, a reading head which can scan one cell of the tape at a time, a finite tape alphabet $S = \{s_0, s_1, \ldots, s_n\}$ of symbols which can be written on the tape, and a finite set $Q = \{q_0, q_1, \ldots, q_m\}$ of possible states. The sets S and Q have the properties that $S \cap Q = \emptyset$, $\{1, B\} \subseteq S$ (where B stands for "blank"), and $q_0 \in Q$ is the designated initial state. A Turing machine which is in state q_j reading symbol s_i on its tape may perform one of three possible actions: it may write over the symbol it is scanning, move the read head right (R), and go into another (possibly the same) state; it may write over the symbol it is scanning, move the read head left (L), and go into another (possibly the same) state; or it may halt.

The action of the Turing machine is governed by a Turing program, given by a transition function δ, whose domain is some subset of $Q \times S$ and whose range is a subset of the set $Q \times S \times \{R, L\}$. If $\delta(q, a) = (\hat{q}, \hat{a}, m)$, then the action of the machine is as follows. If the machine is in state q, reading symbol a on the tape, then it replaces a by $\hat{a}$ on the tape, moves the read head one cell to the right if $m = R$, moves the read head one cell to the left if $m = L$, and goes into state $\hat{q}$. The Turing program halts if the machine is in a state q, reading a symbol a, and the transition function is undefined on (q, a).

A Turing machine computes a partial function as follows: given input $x_1, \ldots, x_n$, the tape is initially set to

$$\ldots B 1^{x_1+1} B 1^{x_2+1} B \ldots B 1^{x_n+1} B \ldots,$$

where 1^k indicates a string of k 1s, one symbol 1 per cell, $\ldots B 1^{x_1+1}$ indicates that all cells to the left of the initial 1 on the tape are blank, and $1^{x_n+1} B \ldots$ indicates that all cells to the right of the last 1 on the tape are blank. The reading head is positioned on the leftmost 1 on the tape, and the machine is set to the initial state q_0. The output of the function (if any) is the number of 1s on the tape when the machine halts, after executing the program, if it ever halts.

The following is a Turing machine program which computes the function $f(x_1, x_2) = x_1 + x_2$, thus showing that f is Turing computable. The idea is that, given input

$$\ldots B 1^{x_1+1} B 1^{x_2+1} B \ldots,$$

the machine replaces the middle blank B by a 1 (instructions $1 - 2$), moves to the leftmost 1 (instructions 3 − 4), and then erases three 1s (instructions 5 − 7).

- $\delta(q_0, 1) = (q_0, 1, R)$

- $\delta(q_0, B) = (q_1, 1, L)$

- $\delta(q_1, 1) = (q_1, 1, L)$

- $\delta(q_1, B) = (q_2, B, R)$

- $\delta(q_2, 1) = (q_3, B, R)$

- $\delta(q_3, 1) = (q_4, B, R)$

- $\delta(q_4, 1) = (q_5, B, R)$

Other formalizations of the Turing machine exist which are slight variations of those given here and which produce the same class of Turing–computable functions.

concentric   A common geometric term, meaning "with the same center". *See* concentric circles, concentric cylinders.

concentric circles   Circles that lie in the same plane and have the same center.

concentric cylinders   Circular cylinders whose circular cross-sections are concentric circles.

concentric spheres   Spheres with the same center.

cone   A solid in $\mathbf{R}^3$, bounded by a region in a plane (the *base*) and the surface formed by straight lines (the *generators*) which join points of the boundary of the base to a fixed point (the *vertex*) not in the plane of the base. The conical surface described by a moving straight line passing through the vertex and tracing any fixed curve, such as a circle, ellipse, etc., at another point is sometimes also called a cone. A *cone* may be viewed as a quadratic surface, whose equation is $Ax^2 + By^2 + Cz^2 = 0$ ($A, B, C \neq 0$). When $A = B$, it is a right circular cone (also called a cone of revolution); if $A \neq B$, it is an oblique circular cone.

cone extension   A deformation of a cone. For a given direction at a point, it represents the increase of length per unit length of arc, i.e., the unit vector in that direction. For example, let I^k ($I = [-1, 1]$) be a k-dimensional convex cell which is a cone of the boundary of I^k from its center 0. Each point x of I^k can be written uniquely as $x = t \cdot u$ for $0 < t \leq 1$ where u belongs to the boundary of I^k. A *cone extension* results when a piecewise linear embedding F from the boundary of I^m to the boundary of I^n is extended to a piecewise linear embedding F between the two convex cells by setting $F(0) = 0$ and $F(t \cdot u) = t \cdot f(u)$ for $t \cdot u \in I^m - \{0\}$.

conformal arc element   Let S^n be a conformal space of dimension n (an n-dimensional sphere represented as the quadric hypersurface S^n: $x_1^2 + x_2^2 + ... + x_n^2 - 2x_0 x_\infty = 0$ in an $(n+1)$-dimensional real projective space P^{n+1}, where the (x_i) are homogeneous coordinates in P^{n+1}). The *conformal arc element* of a curve is given by the Frenet-Serret formula for the curve. For example, let S be a surface in a 3-dimensional projective space and let A be a point of S associated with all the frames $[A, A_1, A_2, A_3]$ where A_1, A_2, A_3 are points of the tangent plane to S at A. The Frenet-Serret formula for S^3 is given by the following matrix:

$$\omega_\alpha^\beta = \begin{bmatrix} 0 & d\alpha & 0 & 0 & 0 \\ \kappa\, d\alpha & 0 & 0 & 0 & d\alpha \\ -d\alpha & 0 & 0 & \tau\, d\alpha & 0 \\ 0 & 0 & -\tau\, d\alpha & 0 & 0 \\ 0 & \kappa\, d\alpha & -d\alpha & 0 & 0 \end{bmatrix}$$

where $d\alpha$ is the conformal arc element, κ is the conformal curvature, and τ is the conformal torsion. ω_α^β is the Pfaffian form that depends on the principal parameters determining the origin A and the secondary parameters determining the frame.

conformal correspondence   A diffeomorphism between two surfaces, whose derivative is a linear map. Angles, but not necessarily lengths, are preserved under *conformal correspondence*. Also called *conformal mapping*.

conformal curvature   Let I be an open interval of $\mathbf{R}$. Let $\alpha : I \to \mathbf{R}^3$ be a curve parameterized by arc length s ($s \in I$) and $\alpha''(s) \neq 0$. For each value of s, let t, n, and b be vector fields along α defined by

$$t(s) = \alpha''(s), \; n(s) = \frac{\alpha''(s)}{\|\alpha''(s)\|}$$

and

$$b(s) = t(s) \times n(s) \; .$$

The derivative $t'(s) = \kappa(s) n(s)$ yields the function $\kappa : I \to \mathbf{R}$, the geometric entity which is the curvature of α in a neighborhood of S. Physically, curvature measures how much the curve differs (bends) from a straight line. This definition is generalizable to n-dimensional conformal space, where the *conformal curvature* of a curve can be derived from the Frenet-Serret apparatus. The concept of curvature associated with a moving frame along a curve was introduced by F. Frenet in 1847 and independently by J.A. Serret in 1851.

conformal differential geometry   The study of geometric quantities that are invariant under conformal transformations, using methods of mathematical analysis such as differential calculus.

conformal equivalence   Let $w = f(z)$ be a function that conformally maps a domain D on the complex z-sphere homeomorphically onto a domain $\triangle$ on the complex w-sphere. Then $\triangle$ is *conformally equivalent* to D.

conformal geometry   The study of properties of figures that are invariant under conformal transformations. Let S^n be an n-dimensional sphere, P^{n+1} be an $(n + 1)$-dimensional projective space, and let $M(n)$ be the group of all projective transformations of P^{n+1} which leave S^n invariant. Then $(S^n, M(n))$ is a *conformal geometry* or a Möbius geometry.

conformal invariant   A geometric quantity preserved by conformal mappings.

conformal mapping   A *conformal mapping* or correspondence between two surfaces S and S^* is a diffeomorphism of S onto S^* such that the angle between any two curves at an arbitrary point x on S is equal to the angle between the corresponding curves on S^*. Conformal mappings are more general than isomorphisms which pre-serve both angles and distances. In $\mathbf{R}^3$, conformal mappings are those obtained by translations, reflections in planes, and inversions in spheres. A one-to-one conformal mapping is a conformal transformation. In $\mathbf{R}^3$ the conformal transformations form the 10-parameter conformal group. In 1779, Lagrange had obtained all the conformal transformations of a portion of the earth's surface onto a plane area that transformed latitude and longitude circles into circular arcs.

conformal torsion   Let I be an open interval of $\mathbf{R}$. Let $\alpha : I \to \mathbf{R}^3$ be a curve parameterized by arc length S ($S \in I$) and $\alpha sf''(s) \neq 0$. For each value of S, let t, n, and b be vector fields along α defined by

$$t(s) = \alpha sf'(s), \; n(s) = \frac{\alpha sf''(s)}{\|\alpha sf''(s)\|},$$

and

$$b(s) = t(s) \times n(s) \; .$$

The derivative $bsf'(s) = \tau(s) n(s)$ yields the function $\tau : I \to \mathbf{R}$, a geometric entity which is the torsion of α in the neighborhood of s. It measures the arc-rate of turning of b. Generalizing to n-dimensional conformal space, the *conformal torsion* of a curve can be derived from the Frenet-Serret apparatus. The concept of torsion associated with a moving frame along a curve was introduced by F. Frenet in 1847 and independently by J.A. Serret in 1851.

congruence of lines   Refers to a set of lines in projective, affine, or Euclidean space depending on a set of parameters. For example, let P^5 be a 5-dimensional projective space, and Q a hyperquadrix defined by the equation $p^{01} p^{23} - p^{02} p^{13} + p^{03} p^{12} = 0$, where the p^{ij} are homogeneous coordinates of P^5. Then we define a congruence of lines as a set of 2-parameter straight lines corresponding to a surface of two dimensions on Q in P^5. The theory of congruences is an important part of projective differential geometry. *See also* congruent objects in space.

congruence on a category   An equivalence relation $\sim$ on the morphisms of a category C such that (i.) for every equivalence class E of morphisms in C, there exist objects A, B in C such that E is contained in the class of morphisms from A to B, and (ii.) for all morphisms f, f', g, g' of C, if $f \sim f'$ and $g \sim g'$, then $f \circ g \sim f' \circ g'$ (assuming $f \circ g$ and $f' \circ g'$ exist).

congruent objects in plane   Two objects P and P^* in $\mathbf{R}^2$ are *congruent* if there exists a rigid motion $\alpha : \mathbf{R}^2 \to \mathbf{R}^2$ such that $\alpha(P) = P^*$.

congruent objects in space   Two objects S and S^* in $\mathbf{R}^3$ are *congruent* if there exists a rigid motion $\alpha : \mathbf{R}^3 \to \mathbf{R}^3$ such that $\alpha(S) = S^*$.

conic section   A geometric locus obtained by taking planar sections of a conical surface (i.e., of a circular cone formed from the rotation of one line around another, provided the lines are not parallel or orthogonal). These sections do not pass through the intersection point of the two lines which produce the circular cone. The *conic section* is thus a plane curve in $\mathbf{R}^3$ generated by a point that moves so that the ratio of its distance from a fixed point to its distance from a fixed line is constant. It can be one of three types: an ellipse (where the intersecting plane meets all generators of the cone in the points of only one convex half-cone), a parabola (where the intersecting plane is parallel to one of the tangent planes to the cone going off to infinity), or a hyperbola (where the intersecting plane meets both half-cones).

conical helix   A space curve that lies on the surface of a cone and cuts all the generators at a constant angle. *See* cone.

conical surface   A surface of revolution of constant curvature in $\mathbf{R}^3$. It can be generated by a straight line that connects a fixed point (the *vertex*) with each point of a fixed curve (the *directrix*). The *conical surface* consists of two concave pieces positioned symmetrically about the vertex. It is *quadric* if the directrix is a conic. A *circular conical surface* is one whose directrix is a circle and whose vertex is on the line perpendicular to the plane of the circle and passing through the center of the circle.

conjugate angles   Two angles whose sum is 360 degrees (2π radians).

conjugate arcs   Two circular arcs whose union is a complete circle.

conjunctive normal form   A propositional (sentential) formula of the form

$$\bigwedge_{i=1}^{n} \left(\bigvee_{j=1}^{m_i} A_{ij} \right) ;$$

i.e.,

$$(A_{11} \vee \cdots \vee A_{1m_1}) \wedge \cdots \wedge (A_{n1} \vee \cdots \vee A_{nm_n}),$$

where each A_{ij}, $1 \le i \le n$, $1 \le j \le m_i$, is either a sentence symbol or the negation of a sentence symbol. Every well-formed propositional formula is logically equivalent to one in *conjunctive normal form*.

For example, if A, B, and C are sentence symbols, then the well-formed formula $((A \to B) \to C)$ is logically equivalent to $(\neg A \vee \neg B \vee C) \wedge (A \vee \neg B \vee C) \wedge (A \vee B \vee C)$, which is a formula in conjunctive normal form.

connected   A subset A of a topological space X is *connected* if it is connected as a subspace of X. In other words, A is connected if there do not exist nonempty disjoint sets U and V, which are relatively open in A such that $A = U \cup V$. That is, there are no disjoint open sets U and V in X with $U \cap A \ne \emptyset$, $V \cap A \ne \emptyset$, and $A = (U \cap A) \cup (V \cap A) = (U \cup V) \cap A$. For example, $(0, 1)$ is connected in $\mathbf{R}$, while $[0, 0.5) \cup (0.5, 1]$ is not, because $[0, 0.5)$ and $(0.5, 1]$ are relatively open in that space.

connected component   In a topological space X, the *connected component* of $x \in X$ is the largest connected $A \subseteq X$ which contains x. Equivalently, A is the union of all connected sets in X which contain x. Any space can be partitioned into its connected components, which must be disjoint.

connected im kleinen   A topological space X is *connected im kleinen* if, for any $x \in X$ and open set U containing x, there is a connected $A \subseteq U$ and an open $V \subseteq A$ with $x \in V$. This form of connectedness is stronger than being locally connected.

connected set   *See* connected.

connected sum   An n-dimensional manifold formed from n-dimensional manifolds M_1 and M_2 (and denoted $M_1 \# M_2$) as follows: Let B_1 and B_2 be closed n-dimensional balls in M_1 and M_2, respectively. Let $h : S_1 \longrightarrow S_2$ be a homeomorphism of the boundary sphere of B_1 to the boundary of B_2. Then $M_1 \# M_2$ is the union of M_1 minus the interior of B_1 and M_2 minus the interior of B_2, with each x in S_1 identified with $h(x)$ in S_2. The resulting space is a manifold; different choices of balls and homeomorphisms may give rise to inequivalent manifolds.

connected topological space   A topological space X such that there do not exist nonempty disjoint open sets U and V such that $X = U \cup V$. Equivalently, the only subsets of X that are both open and closed are $\emptyset$ and X itself.

connective fiber space   A fiber space that cannot be represented as the sum of two nonempty disjoint open-closed subsets. Connectivity is preserved under homeomophisms.

conservative   Let $\mathcal{L}_1$ and $\mathcal{L}_2$ be first order languages with $\mathcal{L}_1 \subseteq \mathcal{L}_2$; let T_1 be a theory of $\mathcal{L}_1$ and T_2 be a theory of $\mathcal{L}_2$. The theory T_2 is a *conservative extension* of T_1 if

(i.) T_2 is an extension of T_1; i.e., $T_1 \subseteq T_2$, and

(ii.) for every sentence σ of $\mathcal{L}_1$, if σ is a theorem of T_2 ($T_2 \vdash \sigma$), then σ is a theorem of T_1 ($T_1 \vdash \sigma$).

consistent   Let $\mathcal{L}$ be a first order language and let Γ be a set of well-formed formulas of $\mathcal{L}$. The set Γ is *consistent* if it is not inconsistent; i.e., if there does not exist a formula α such that both α and $(\neg \alpha)$ are theorems of Γ. If Γ is consistent, then there must exist a formula α such that α is not a theorem of Γ. Note that Γ is consistent if and only if Γ is satisfiable, by soundness and completeness of first order logic.

If Γ is a set of sentences and φ is a well-formed formula, then φ is consistent with Γ if Γ has a model that is also a model of φ.

consistent axioms   A set of axioms such that there is no statement A such that both A and its negation are provable from the set of axioms. Informally, a collection of axioms is consistent if there is a model for the axioms; the axioms of group theory are consistent as $G = \{e\}$, where $\cdot$ is defined on G by $e \cdot e = e$, is a model of these axioms.

In the case of formal systems, Gödel's Second Incompleteness Theorem states, roughly, that the consistency of any sufficiently strong theory cannot be proved in that theory; for example, it is not possible to prove the consistency of Zermelo-Fraenkel (ZF) set theory from the axioms of ZF. Consequently, only a relative notion of consistency can be considered; i.e., given a set Σ of axioms of a formal system and a statement A in the language of that system, one asks whether $\Sigma \cup \{A\}$ is consistent, assuming that Σ is consistent. For example, Gödel proved in 1936 that, assuming ZF is consistent, so is ZFC, where ZFC is ZF set theory together with the Axiom of Choice. In addition, Cohen proved in 1963 that, assuming ZF is consistent, so is $ZF + \neg AC$, where $ZF + \neg AC$ is ZF set theory together with the negation of the Axiom of Choice.

constructible set   An element of the class L, defined below in the constructible hierarchy:

(i.) $L_0 = \emptyset$;

(ii.) $L_\alpha = \bigcup_{\beta < \alpha} L_\beta$, if α is a limit ordinal;

(iii.) $L_{\alpha+1}$ = the set of all subsets definable over L_α;

(iv.) $L = \bigcup_{\alpha \in \mathbf{Ord}} L_\alpha$.

contact element   In the Euclidean plane $\mathbf{R}^2$, an ordered pair (p, ℓ) consisting of a point p and a line ℓ containing the point p. More generally, a *contact element* in a smooth n-dimensional manifold M is a pair (p, H) consisting of a point p in M and an $n - 1$-dimensional plane H in the tangent space at p. The contact elements in M form a $(2n-1)$-dimensional manifold which has a special structure on it called a contact structure.

contact form   A one-form ω on a smooth $(2n + 1)$-dimensional manifold M such that $\omega \wedge (d\omega)^n \neq 0$ everywhere on M. At each point p in M, the kernel $K(p)$ of $\omega(p)$ is a plane of dimension $2n$ in the tangent space at p. The condition on ω is equivalent to the statement: for any vector $V \in K(p)$ there is a W such that $d\omega(V, W) \neq 0$. Darboux's Theorem says that it is always possible to find local coordinates $(x^1, \ldots, x^n, y^1, \ldots, y^n, z)$ in which $\omega = dz - \sum_{i=1}^{n} y^i dx^i$.

contact manifold   The odd-dimensional counterpart of a symplectic manifold. A *contact manifold* is a smooth $(2n+1)$-dimensional manifold M together with a one-form ω on M such that $\omega \wedge (d\omega)^n \neq 0$ everywhere on M. There is a unique vector field V on M, called the characteristic vector field, determined by two conditions:

(i.) $\omega(V) = 1$ and

(ii.) $d\omega(V, W) = 0$ for all W.

In special local coordinates, this is the vector field $\frac{\partial}{\partial z}$.

An example of a contact manifold is the unit sphere S^{2n+1}, viewed as a submanifold of complex n-dimensional space. The characteristic vector field is the field of unit vectors tangent to the great circles which form the fibers of the Hopf fibration $\pi : S^{2n+1} \longrightarrow \mathbf{CP}^n$ of the sphere over complex projective space. π is the map that takes a point in the sphere to the complex 1-dimensional subspace containing the point. The 1-form ω is given by the formula

$$\omega(W) = \langle V, W \rangle ,$$

where $\langle , \rangle$ is the Euclidean inner product.

contact metric structure   A Riemannian metric g on a manifold M of dimension $2n + 1$ and a contact form η which are compatible. The contact form η is a 1-form on M which satisfies the condition $\eta \wedge (d\eta)^n \neq 0$ at every point. η determines a subspace $K(p)$ of codimension 1 in the tangent space M_p to M at the point p, as well as a vector ξ transverse to $K(p)$ and determined by the conditions: $\eta(\xi) = 1$ and $d\eta(V, \xi) = 0$ for all V in $K(p)$. The metric g must make $\xi(p)$ a unit vector, orthogonal to $K(p)$. For X and Y in $K(p)$, the metric satisfies $d\eta(X, Y) = g(X, \phi Y)$, where $\phi(p) : M_p \to M_p$ is a linear transformation satisfying $\phi^2(V) = -V$ on $K(p)$ and $\phi(\xi) = 0$. More precisely, ϕ is a tensor field on M of type $(1, 1)$ satisfying $\phi^2 = -I + \eta \otimes \xi$.

contact structure   A specification of a $(2n)$-dimensional plane $K(p)$ in the tangent space of a manifold M of dimension $2n + 1$, at each point p, in such a way that in some open set U around p there is a smooth, one-form ω such that $K(q)$ is the kernel of $\omega(q)$ for every q. The one-form is required to satisfy the non-degeneracy condition: $\omega \wedge (d\omega)^n \neq 0$ for all q in U. The form ω is called a local contact form. On the overlap of two such open sets, the local forms agree up to a non-vanishing scalar multiple. If the form can be globally defined, then M is a contact manifold.

An example of a *contact structure* is given by the manifold of straight lines in the plane $\mathbf{R}^2$. Local coordinates (x, y, z) are given by letting (x, y) be the point of contact and $0 < z < \pi$ the angle between the line and a horizontal line. (Different coordinates are chosen if the line is horizontal.) The 1-form is given by $\omega(x, y, z) = \sin(z)dx - \cos(z)dy$. This 1-form cannot be defined globally.

continued fraction   A real number of the form

$$a_0 + \cfrac{b_1}{a_1 + \cfrac{b_2}{a_2 + \cfrac{b_3}{a_3 + \cfrac{b_4}{a_4 + \ddots}}}}$$

where each a_i is a real number. If the expression consists of only a finite number of fractions, the expression is called a *finite* continued fraction and is (obviously) a rational number. *See also* finite continued fraction. An *infinite simple* continued fraction is one of the form

$$a_0 + \cfrac{1}{a_1 + \cfrac{1}{a_2 + \cfrac{1}{a_3 + \ddots}}}$$

where each of the a_i is an integer. *See also* convergence of a continued fraction.

continuous function   A function f from a topological space X to a topological space Y such that, for each open $U \subseteq Y$, the set $f^{-1}(U)$ is open in X. For functions on the real line, this is equivalent to requiring that, for any $\epsilon > 0$, there exists a $\delta > 0$ such that $|f(x) - f(y)| < \epsilon$ whenever $|x - y| < \delta$.

continuous geometry An orthomodular lattice, i.e., a complete and complemented modular lattice L such that given any element x of L and any subset W of L which is well ordered with respect to the ordering in L, then $x \cap \sup w = \sup(a \cap w)$, where $w \in W$. The concept of a *continuous geometry* was introduced by John von Neumann. When the dimension is discrete, continuous geometry contains projective geometry as a special case. Generally, however the lattices have continuous dimension.

continuum hypothesis The statement

$$2^{\aleph_0} = \aleph_1 .$$

(It is not possible to prove this statement or its negation in Zermelo-Frankel set theory with the Axiom of Choice.)

contractible topological space A topological space X that can be shrunk to a point. More precisely, X is contractible if there is a continuous function $c : X \times [0, 1] \longrightarrow X$ (called a *contraction*) such that $c(x, 0) = x$ for all x in X, and for some p in X, $c(x, 1) = p$ for all x. The n-dimensional Euclidean spaces are contractible, while any space with a nonzero homology or homotopy group in a positive dimension gives an example of a non-contractible space.

contraction If X is a subspace of Y, then a *contraction* of X in Y is a continuous function $c : X \times [0, 1] \longrightarrow Y$ such that $c(x, 0) = x$ for all x in X, and for some p in Y, $c(x, 1) = p$ for all x in X. *See* contractible topological space.

convergence of a continued fraction Given the continued fraction

$$a_0 + \cfrac{b_1}{a_1 + \cfrac{b_2}{a_2 + \cfrac{b_3}{a_3 + \cfrac{b_4}{a_4 + \cdots}}}}$$

define

$$C_n = a_0 + \cfrac{b_1}{a_1 + \cfrac{b_2}{a_2 + \cfrac{b_3}{a_3 + \cfrac{\ddots}{a_{n-1} + \frac{b_n}{a_n}}}}}$$

(the nth *convergent* of the continued fraction). If $\lim_{n \to \infty} C_n = L$, then the continued fraction is said to converge to L. That is,

$$L = a_0 + \cfrac{b_1}{a_1 + \cfrac{b_2}{a_2 + \cfrac{b_3}{a_3 + \cfrac{b_4}{a_4 + \cdots}}}}$$

convergent *See* convergence of a continued fraction.

convergent sequence A sequence of points $\{x_n : n \in \mathbf{N}\}$ in a topological space X *converges to* $x \in X$ if, for any open set U containing x, there is an $N \in \mathbf{N}$ with $x_n \in U$ for all $n \geq N$. A sequence is *convergent* if it converges to some $x \in X$. For example, the sequence of reals $\{\frac{1}{n}\}$ converges to 0, while the sequence $\{n\}$ does not converge.

convex (1) A non-empty subset X of $\mathbf{R}^n$ such that, for any elements $x, y \in X$, and any number c such that $0 \leq c \leq 1$, the element $cx + (1-c)y$ of $\mathbf{R}^n$ belongs to X. In $\mathbf{R}^2$, for example, a set is *convex* if it contains the line segment joining any two of its points. *See also* convex closure.
(2) A real function $f(x)$ in an interval I such that the graph of f lies nowhere above its secant line in any subinterval of I.

convex body A bounded, closed, convex set (finite or infinite) that has interior points. *See* convex.

convex cell The convex closure of a finite set of points $P = \{p_0, p_1, ..., p_k\}$, in an n-dimensional affine space A^n. When $p_0, ..., p_k$ are independent, it is a k-dimensional simplex with vertices $p_0, ..., p_k$.

convex closure For any subset X of an n-dimensional affine space A^n, there exists a minimal convex set that contains X. This set, which is the intersection of all the convex sets that contain X, is the *convex closure* of X. In $\mathbf{R}^n$, the convex closure of a set X is the set of possible locations of the center of gravity of mass which can be distributed in different ways in the minimal convex set containing X. Each point of the convex closure is the center of gravity of a mass concentrated at not more than $n + 1$ points.

For a subset X of A^n, X is *convex* if the segment joining two arbitrary points of X is contained in X.

convex cone   A convex body consisting of half-lines emanating from a point (the *apex* of the cone). The surface of a *convex cone* is sometimes called a convex cone.

convex cylinder   A cylinder that lies entirely on one side of any tangent plane at a point of the cylinder. *See* cylinder.

convex hull   The smallest convex set containing a given subset X of a Euclidean space. The *convex hull* of X can be constructed by forming the intersection of all half-planes containing X. *See also* convex closure.

convex polygon   A polygon in $\mathbf{R}^2$ with the property that each of its interior angles (the angles made by adjacent sides of the polygon and contained within the polygon) is less than or equal to 180 degrees. A *convex polygon* always has an interior.

convex polyhedral cone   A convex cone in $\mathbf{R}^3$ which is the intersection of linear half-spaces. *See* convex cone.

convex polyhedron   The convex closure of a finite number of points in $\mathbf{R}^n$; that is, the bounded intersection of a finite number of closed half-spaces. In $\mathbf{R}^3$, it is a solid bounded by plane polygons, which lies entirely on one side of any plane containing one of its faces. Any plane section of a *convex polyhedron* is a convex polygon. *See* convex closure, convex polygon.

coordinate   *See* coordinate system. In $\mathbf{R}^2$, a *coordinate* is one of an ordered pair of numbers that locates the position of a point in the plane.

coordinate axis   One of finitely many components of a reference system which provides a one-to-one correspondence between the elements of a set (on a plane or a surface, in a space, or on a manifold) with the numbers used to specify their position. In $\mathbf{R}^n$, it is part of an orthogonal frame which determines the rectangular coordinates of each point in the space. In $\mathbf{R}^2$ a *coordinate axis* is a line along which or parallel to which a coordinate is measured.

coordinate bundle   Let E, B, F be topological spaces and $p : E \to B$ be a continuous map. Let G be an effective left topological transformation group of F. If there exists an open covering $\{U_\alpha\}_{\alpha \in \Lambda}$ of B, and a homeomorphism $\phi_\alpha : U_\alpha \times F \approx p^{-1}(U_\alpha)$ for each $\alpha \in \Lambda$, then the system $(E, p, B, F, G, U_\alpha, \phi_\alpha)$ is a coordinate bundle if it has the following three properties: (i.) $p\phi_\alpha(b, y) = b$ ($b \in U_\alpha$, $y \in F$); (ii.) define $\phi_{\alpha, \beta} : F \approx p^{-1}(b)$ ($b \in U\alpha$) by $\phi_{\alpha,\beta}(y) = \phi_\alpha(b, y)$. Then $g_{\beta\alpha}(b) = \phi_{\beta,b}^{-1}\phi_{\alpha,b} \in G$ for $b \in U_\alpha \cap U_\beta$; (iii.)$g_{\beta\alpha}(b) : U_\alpha \cap U_\beta \to G$ is continuous.

coordinate function   The homeomorphism

$$\phi_\alpha : U_\alpha \times F \approx p^{-1}(U_\alpha)$$

for each $\alpha \in \Lambda$ of the coordinate bundle $(E, p, B, F, G, U_\alpha, \phi_\alpha)$ belonging to the fiber bundle (E, p, B, F, G).

coordinate hyperplane   A *coordinate hyperplane* in a vector space X over a field K is the image under a translation of a vector subspace M with the quotient space X/M.

coordinate neighborhood   The open covering $\{U_\alpha\}(\alpha \in \Lambda)$ of B of the coordinate bundle $(E, p, B, F, G, U_\alpha, \phi_\alpha)$ belonging to the fiber bundle (E, p, B, F, G). *See* coordinate bundle.

coordinate system   Let S be a set of mathematical objects. A *coordinate system* is a mechanism that assigns (tuples of) numbers to each element of the set S. The numbers corresponding to each element are called its *coordinates*. In $\mathbf{R}^2$, such a reference system is called the rectangular coordinate system. In $\mathbf{R}^3$, the coordinate functions, u and v, of p^{-1} [where p is a one-to-one regular mapping of an open set of $\mathbf{R}^2$ into a subset of $\mathbf{R}^3$] constitute the coordinate system associated with p.

coordinate transformation   Let E, B, F be topological spaces and $p : E \to B$ be a continuous map. Let G be an effective left topological transformation group of F. Assume there exists an open covering $\{U_\alpha\}$ ($\alpha \in \Lambda$) of B, and a homeomorphism $\phi_\alpha : U_\alpha X F \approx p^{-1}(U_\alpha)$ for each $\alpha \in \Lambda$. Define $\phi_{\alpha,\beta} : F \approx p^{-1}(b)$ ($b \in U_\alpha$) by $\phi_{\alpha,\beta}(y) = \phi_\alpha(b, y)$. Then $g_{\beta\alpha}(b) = \phi_{\beta,b}^{-1}\phi_{\alpha,b} \in G$ for $b \in U_\alpha \cap U_\beta$. Then the continuous transformation $g_{\beta\alpha}$ is the *coordinate transformation* of the coordinate bundle $(E, p, B, F, G, U_\alpha, \phi_\alpha)$ belonging to the fiber bundle (E, p, B, F, G).

coplanar   Lying in the same plane.

coproduct   For any two sets X and Y, a *coproduct* of X and Y is a disjoint union D in the diagram

$$X \xrightarrow{i} D \xleftarrow{j} Y,$$

where i and j are injections. The set D is not unique and can be constructed as follows. If X and Y are disjoint, then let $D = X \cup Y$. If X and Y are not disjoint, let $D = X \cup Y'$, where Y' is a set that is equivalent to Y and disjoint from X. Other coproducts of X and Y can be formed by choosing different sets Y'.

corresponding angles   Let two straight lines lying in $\mathbf{R}^2$ be cut by a transversal, so that angles x and y are a pair of alternating interior angles, and y and z are a pair of vertical angles. Then x and z are *corresponding angles*.

cotangent bundle   Let M be an n-dimensional differentiable manifold of class C^r. Consider $T(M)$, the union over $p \in M$ of all the vector spaces $T_p(M)$ of vectors tangent to M at p. Define $\pi : T(M) \to M$ by $\pi(T_p(M)) = p$. Then $T(M)$ may be regarded as a manifold called the *tangent bundle* of M. The dual bundle is the *cotangent bundle*.

coterminal angles   Angles with the same terminal side (the moving straight line which revolves around the fixed straight line, the initial side, to form the angle) and the same initial side. Two angles are *coterminal* if they are generated by the revolution of two lines about the same point of the initial side in such a way that the final positions of the revolving lines are the same. For example, 60° and -300° are *coterminal angles*.

Countable Axiom of Choice   The statement that, for any countable family of non-empty, pairwise disjoint sets $\{X_\alpha\}_{\alpha \in \Lambda}$, there exists a set Y which consists of exactly one element from each set in the family. Equivalently, if $\{X_\alpha\}_{\alpha \in \Lambda}$ is a countable family of non-empty sets, then there exists a function $f : \{X_\alpha\}_{\alpha \in \Lambda} \to \bigcup_{\alpha \in \Lambda} X_\alpha$ such that $f(X_\alpha) \in X_\alpha$ for all $\alpha \in \Lambda$. See Axiom of Choice.

countable chain condition   (1) A partial order $(\mathcal{P}, \leq)$ has the *countable chain condition* if any antichain in $\mathcal{P}$ is countable. A set $A \subseteq \mathcal{P}$ is an antichain if its elements are pairwise incompatible; that is, for any p and q in A, there is no $r \in \mathcal{P}$ with $r \leq p$ and $r \leq q$. Thus, $\mathcal{P}$ has the countable chain condition (or is ccc) if it has no uncountable subset of pairwise incompatible elements. Examples of ccc partial orders include the collection of all finite sequences of 0s and 1s ordered by extension ($p \leq q$ if $p \supseteq q$), and the collection of all measurable sets modulo measure zero sets ordered by inclusion ($[A] \leq [B]$ if $A \subseteq B$).

(2) A topological space satisfies the *countable chain condition* if it contains no uncountable collection of pairwise disjoint non-empty open sets. If X is a topological space and $\mathcal{P}$ is the collection of all non-empty open subsets of X ordered by $p \leq q$ if and only if $p \subseteq q$, then X is a ccc topological space if and only if $\mathcal{P}$ is a ccc partial order.

countably compact topological space   A topological space X such that any countable open cover of X contains a finite subcover. That is, if each U_n is open and $X = \bigcup_{n \in \mathbf{N}} U_n$, then there is a finite set $A \subseteq \mathbf{N}$ with $X = \bigcup_{n \in A} U_n$. This is equivalent to requiring that any countably infinite subset of X has an accumulation point.

The space of ordinals less than ω_1 with the order topology is a countably compact space which is not compact.

counterclockwise   The direction of rotation opposite to that in which the hands of the clock move.

covariant functor   Let $\mathcal{C}$ and $\mathcal{D}$ be categories. A *covariant functor* F is a function $F : \mathrm{Obj}(\mathcal{C})$

$\rightarrow$ Obj($\mathcal{D}$) such that, for any A, B $\in$Obj($\mathcal{C}$) the following hold:

(i.) if $f \in \text{Hom}_{\mathcal{C}}(A, B)$, then

$$F(f) \in \text{Hom}_{\mathcal{D}}(F(A), F(B)) ;$$

(ii.) $F(1_{\mathcal{C}}) = 1_{\mathcal{D}}$;

(iii.) $F(gf) = F(g)F(f)$, where f and g are morphisms in $\mathcal{C}$ whose composition gf is defined.

covering dimension   A nonnegative integer, assigned to a set by means of coverings. For topological spaces the *covering dimension* (or Lebesque dimension) is defined in terms of open coverings. The dimension of a normal space x is less than or equal to n if, in each finite open covering of x, a finite open covering can be inscribed whose number of elements containing a given point is less than or equal to $n + 1$.

covering group $p: E \longrightarrow X$ is a covering space of X if every point of X has a neighborhood whose inverse image is the disjoint union of open sets homeomorphic to the neighborhood by p. The *covering group* (group of covering transformations) is the group of homeomorphisms of E which preserve the fibers (homeomorphisms $h : E \longrightarrow E$ with $ph = p$).

The real line covers the circle S^1 by the map which takes x to $e^{2\pi i x}$. The group of covering transformations is isomorphic to the group of integers.

covering homotopy property   A map $p: E \longrightarrow B$ has the *covering homotopy property* if it satisfies the following: given any map $f: X \longrightarrow E$ and any homotopy $h: X \times [0, 1] \longrightarrow B$ of $p \circ f$ (so $hi = pf$), there is a lift $H: X \times [0, 1] \longrightarrow E$ of h so that $pH = h$ and $Hi = f$, where i is the inclusion of $X \times \{0\}$ in $X \times [0, 1]$.

That is, given maps represented by the solid lines in the diagram below,

$$
\begin{array}{ccc}
X & \xrightarrow{f} & E \\
i \downarrow & & \downarrow p \\
X \times I & \xrightarrow{h} & B
\end{array}
$$

there exists a map $H : X \times I \longrightarrow E$ so that the diagram with H added commutes.

A surjective map satisfying the covering homotopy property is called a fibration. This is a generalization of the concept of fiber bundle; the fiber in a fibration is only determined up to homotopy, that is, the inverse images of two different points (if B is connected) will only be homotopy equivalent to each other.

covering map   A continuous surjection $p :$ $X \rightarrow Y$ such that every $y \in Y$ has an open path connected neighborhood U such that, for each open path component $V \subseteq p^{-1}(U)$, p is a homeomorphism from V onto U. In other words, the Vs form a stack of copies over U which cover it. For example, the map $p(t) =$ $(\cos t, \sin t)$ is a *covering map* from $\mathbf{R}$ to the unit circle.

covering space   A topological space X is a *covering space* of Y if there is a covering map $p : X \rightarrow Y$. *See* covering map. For example, $\mathbf{R}$ is a covering space of the unit circle via $p(t) = (\cos t, \sin t)$.

covering transformation   A map $\phi : E \longrightarrow E$ such that $p \circ \phi = \phi$, where $p : E \longrightarrow B$ is a covering map of an arcwise connected, locally arcwise connected space B. *See* covering map.

covering transformation group   The group of covering transformations $\phi : E \longrightarrow E$, under composition. *See* covering transformation. In the special case where E is the universal cover of B, this group is isomorphic to the fundamental group of the space B.

cross-section   A *cross-section* or section of a fiber bundle $p : E \longrightarrow B$ with fiber F is a map $s : B \longrightarrow E$ with ph equal to the identity on B. Clearly every trivial bundle $B \times F$ has numerous sections. A non-trivial example is the Möbius band, a bundle over its middle circle. The inclusion of the middle circle is a section for that bundle.

cube   One of the five types of regular polyhedra in E^3. Also known as a hexahedron, it is a solid bounded by 6 planes with 12 equal edges and face angles that are right angles. In E^n, it is a set consisting of all those points $x = (x_1, ..., x_n)$, for which x_i is such that $a_i \leq x_i \leq b_i$ for each i, where a_i and b_i are such that $b_i - a_i$ has the same value for each i.

cumulative hierarchy   The hierarchy of sets defined recursively using the power set operation at successor stages and union at limit stages: (i.) $V_0 = \emptyset$, (ii.) $V_{\alpha+1} = \mathcal{P}(V_\alpha)$, for all ordinals α, and (iii.) $V_\lambda = \bigcup_{\beta<\lambda} V_\beta$, for any limit ordinal λ. Also known as the *Zermelo hierarchy*. *See also* universe of sets.

curvature   A measure of the quantitative characteristics (in terms of numbers, vectors, tensors) which describe the degree to which some object (a curve, Riemannian manifold, etc.) deviates from certain other objects (a straight line, a Euclidean surface, etc.), which are considered to be flat. As a local property of a plane curve, *curvature* may intuitively be thought of as the degree to which a curve is "bent" at each point. For non-planar space curves, curvature is defined as the magnitude of a rate of rotation vector. Gauss had defined the curvature of a surface in $\mathbf{R}^3$ at a point (x, y, z) as the limit of the ratio of the area of the region on a unit sphere around a point (X, Y, Z) [determined by the radius of the sphere in a direction normal to (x, y, z)] to the area of the region on the surface around (x, y, z), as these two areas shrink to their respective points. Riemann's conception of curvature for any n-dimensional manifold was a generalization of Gauss' definition for surfaces.

curve   A continuous function from (an interval in) $\mathbf{R}$ into $\mathbf{R}^n$, although usually referred to as the image (range) of such a function.

Euclid distinguished lines from *curves,* but today lines, in the Euclidean sense, are considered curves which include straight lines. Georg Cantor defined a curve as a continuum that is nowhere dense in $\mathbf{R}^2$. A continuous curve in $\mathbf{R}^2$ that covers a square is a *Peano curve.*

curve of constant breadth   Let E be the boundary of a convex body X in $\mathbf{R}^2$, O an interior point of X, and P an arbitrary point of X different from O. E admits exactly one supporting line $l(P)$ which is perpendicular to the line OP and meets the half-line OP. Let OP' be the half-line with direction opposite to that of OP, and l' the supporting line $l(P')$. Then the distance between the parallel lines l and l' is the *breadth* of E. Let $M = M(E)$ be the maximum

and $m = m(E)$ be the minimum of the breadth of E. If $M = m$, then E is a *curve of constant breadth.*

cusp   A double point on a curve C, at which two tangents to C are coincident.

cut point   (1) In topology, a point p in a space X such that $X\backslash\{p\} = A \cup B$, where A and B are nonempty open sets.

(2) In Riemannian geometry, if M is a Riemannian manifold and p is a point of M, then a point q of M is a *cut point* with respect to p if there is a shortest geodesic joining p to q which, if extended beyond q, fails to be a shortest path to points beyond q. For example, on the standard sphere the antipode of any point is the unique cut point.

CW complex   A topological space constructed iteratively as follows. Let D^n be an n-cell, that is, a set homeomorphic to all points of distance at most 1 from the origin in $\mathbf{R}^n$; the boundary of D^n is the $(n-1)$-sphere, S^{n-1}. Begin with a collection of points. At each stage, attach new cells by identifying the boundary of a cell D^n with points in lower dimensional cells.

One can build the sphere S^n in two distinct ways. First, start with one point; attach the cell D^n by identifying all points in its boundary with that one point. Alternately, start with two points. Attach two 1-cells (intervals) by sending one endpoint to each of the two points that formed the previous stage. This constructs a circle, S^1. Given S^{n-1} by iteration, form S^n by identifying the boundaries of each of two n-cells with S^{n-1}. This latter construction is convenient for studying real projective space.

Any manifold can be constructed as a *CW complex* and such constructions are useful for calculating the homology or cohomology of the manifold.

cycle   An element c in C_n such that $\partial_n(c) = 0$, where C_n is a member of a chain complex over a ring R. *See* chain complex.

cycle group   If $\{C_n, \partial_n\}$ is a chain complex (of Abelian groups), then the kth *cycle group* Z_k is the subgroup of C_k consisting of all elements c such that $\partial_k(c) = 0$. The *boundary group*

$B_n = \partial_{n+1}/C_{n+1}$ is a subgroup of the *cycle group,* since $\partial_n + 1 \circ \partial_n = 0$. *See* chain complex.

cylinder   A flat ruled surface in $\mathbf{R}^3$, which is the locus of points on a straight line (the *generator*) moving parallel to itself, and intersecting a given plane curve (the *directrix*). Thus, as the generator, which is perpendicular to the plane of the directrix, moves along the directrix, it sweeps out a *cylinder.*

cylinder of revolution   A surface formed by the set of all lines passing through a given circle and perpendicular to the plane of the circle. Also known as a *right circular cylinder.*

cylindrical helix   A helix whose path lies on a cylinder, forming a constant angle with the elements of the cylinder. A *cylindrical helix* is described, for example, by the parametric equations

$$x = \cos t$$

$$y = \sin t$$

$$z = t$$

$(-\infty < t < \infty)$.

cylindrical surface   A surface generated by a line in space moving along a curve, always staying parallel to a fixed line.

D

Darboux's frame   If $\gamma(s)$ is a curve on a surface M in Euclidean space $\mathbf{R}^3$, parameterized by arc length, then there is an orthonormal frame (T, U, V) defined along the curve called *Darboux's frame*. The first vector, $T = \gamma'(s)$, is the unit tangent vector to the curve. The vector $V = \nu(\gamma(s))$ is the unit normal to the surface at the point. (This assumes that a normal vector field $\nu(P)$ is specified.) The vector $U = V \times T$ is the normal vector to the curve in the surface, whose direction is determined by the choice of surface normal. One can then define the analog of the Frenet equations for the curve:

$$
\begin{aligned}
T' &= && \kappa_g U &+ \kappa_n V \\
U' &= -\kappa_g T && &+ \tau_g V \\
V' &= -\kappa_n T && - \tau_g U &
\end{aligned}
$$

The functions κ_g and κ_n are the *geodesic curvature* and the *normal curvature*, respectively, of the curve in the surface. τ_g is the *geodesic torsion*.

decagon   A ten-sided polygon.

decidable   A set of objects of some sort is *decidable* if there is an effective procedure (algorithm) which, given an arbitrary object, decides whether or not the object is in the set. This notion, as defined, is an intuitive notion. To define this notion formally requires the formal notion of computability and the ability to code effectively (Gödel number) the objects in question. *See* computable.

For example, the set of tautologies of propositional logic is decidable; i.e., there is an effective procedure which, given a string from the language of the propositional logic, can determine whether or not the string is a well-formed formula, and if it is, can determine whether or not the well-formed formula is a tautology. As another example, a theory in first order logic is decidable if it is both complete and axiomatizable by a decidable set of axioms.

decision problem   A *decision problem* asks if there exists an effective procedure (algorithm) for deciding the truth or falsity of an entire class of statements. If such an effective procedure exists, the problem is said to be *solvable*; otherwise the problem is said to be *unsolvable*.

This notion, as defined, is intuitive. To define it formally requires the formal notion of computability, the ability to effectively code (Gödel number) the statements, and the Church–Turing Thesis.

For example, given an n-ary predicate R, the decision problem associated with R asks if there is an effective procedure for deciding, given arbitrary natural numbers $a_1, \ldots, a_n$, whether $R(a_1, \ldots, a_n)$ is true or false. This decision problem is solvable if R is recursive (computable) as a relation; otherwise it is unsolvable.

There are many famous decision problems. Hilbert's Tenth Problem asks if there is an effective procedure which, given a Diophantine equation $p(x_1, \ldots, x_n) = 0$, where p is a polynomial with integer coefficients, determines whether or not there is an integer solution. It was proved by results of Matijasevič (1970) and Davis, Putnam, and Robinson (1961) that Hilbert's Tenth Problem is not solvable. The decision problem for propositional logic, which asks if there is an effective procedure which, given a well-formed formula of propositional logic, determines whether or not that formula is provable, was proved by Post (1921) to be solvable. On the other hand, the decision problem (Entscheidungsproblem) for first-order logic is not solvable (Church, 1936; Turing, 1936); i.e., if $\mathcal{L}$ is a first-order language with a k-ary function or predicate symbol, for $k \geq 2$, then there is no effective procedure that can determine whether or not an arbitrary sentence of $\mathcal{L}$ is logically valid.

Dedekind cut   Suppose that X is a subset of $\mathbf{R}$ and A, B are two subsets of X with the following properties:

(i.) $A \cap B = \emptyset$;

(ii.) $A \cup B = X$;

(iii.) $a < b$ for any $a \in A$, $b \in B$.

The sets A and B form a *Dedekind cut* if A has a last element and B has no first element, or if A has no last element and B has a first element. X is an interval of $\mathbf{R}$ if and only if any choice of

A, B satisfying conditions (i.) through (iii.) is a Dedekind cut.

If there exists a choice of A and B which is not a Dedekind cut, then X can be extended to an interval of $\mathbf{R}$ by including additional real numbers. For example, if X consists of the rational numbers, then no choice of A, B is a Dedekind cut. However, X may be extended by including the irrational numbers. This technique can be used if X is a subset of any complete linearly ordered set.

deduction   *See* proof.

deficient number   A positive integer n having the property that the sum of its positive divisors is less than $2n$, i.e., $\sigma(n) < 2n$. For example, 16 is deficient, since

$$1 + 2 + 4 + 8 + 16 = 31 < 32 .$$

Compare with abundant number, perfect number.

definition by recursion   *See* recursion.

degenerate conic   A conic section formed by a plane intersecting a conical surface either only in the vertex of the surface (resulting in a point), only at an element of the surface (resulting in a line), or only at two elements of the surface (resulting in two lines).

degenerate simplex   An n-simplex (map from the n-dimensional analog of a triangle to a space) whose image is less than n-dimensional.

degree of an arc   The degree measure of the central angle determining the arc, provided the arc is minor. If the arc is major, its degree is 360 minus the degree measure of the central angle.

degree of mapping   Let $f : (S^n, x_0) \to (S^n, x_0)$ be a continuous map where $n \geq 1$ and let $f_* : H_n(S^n, x_0) \to H_n(S^n, x_0)$ be the induced map. If z_0 is the generator of $H_n(S^n, x_0)$, then $f_*(z_0) = d \cdot z_0$ for some integer d. In this case d is the *degree* of the map f and is denoted $\deg(f)$.

An important consequence of this is the *Brouwer Fixed-Point Theorem:* Every continuous map $f : B^n \to B^n$ has a fixed point.

degree of unsolvability   For a set of natural numbers of A, the set

$$\deg(A) = \{B : B \equiv_T A\};$$

i.e., the class of all sets B of natural numbers which are Turing equivalent to A. A *degree of unsolvability* is often called a Turing degree, or simply a degree.

As an example, if A is a recursive (computable) set, then $\deg(A)$ consists of all recursive (computable) sets. However, if A is computably (recursively) enumerable, then $\deg(A)$ does not consist of all computably enumerable sets.

de Morgan's laws   Let B be a Boolean algebra with binary operations $\cup$, $\cap$ and unary operation $'$. If $X, Y \in B$, then:

$$(X \cap Y)' = X' \cup Y'$$

$$(X \cup Y)' = X' \cap Y' .$$

See Boolean algebra.

denominator   The number b in the fraction $\frac{a}{b}$.

dense linear ordering   Given a set A with at least two distinct elements and a linear ordering $\leq$ on A, $\leq$ is *dense* if, for all $x, y \in A$ with $x < y$, there exists $z \in A$ with $x < z < y$. The usual ordering $\leq$ on $\mathbf{Q}$, the set of rational numbers, is dense.

dense subset   A subset A of a topological space X such that $\bar{A} = X$, where $\bar{A}$ denotes the closure of A in X.

denumerable set   A set A, equinumerous with the set $\mathbf{N}$ of natural numbers; i.e., such that there is a bijection from A onto $\mathbf{N}$. For example, the set $\mathbf{Q}$ of rational numbers is denumerable, while the set $\mathbf{R}$ of real numbers is not denumerable.

derived set   The set of accumulation points of a subset A of a topological space X. The *derived set* of A is written A'. The closure of A is then $A \cup A'$, and A is closed if and only if $A' \subseteq A$.

determined   Let X be a subset of ω^ω. Associated with X is the game G_X which is played by two players A and B. The two players alternate choosing natural numbers $a_1, b_1, a_2, b_2, \ldots$. If the sequence $(a_1, b_1, a_2, b_2, \ldots) \in X$, then player A wins. If not, player B wins. A strategy is a rule that tells the player (either A or B) which number to choose, based on the previous choices of both players. A winning strategy is a strategy in which the player always wins. The game G_X is *determined* if one of the two players has a winning strategy.

developable surface A ruled surface is a surface swept out by a straight line (called the generator) moving through space. It can be given a parameterization of the form $X(u, v) = C(u) + v D(u)$ where C is a curve and $C'(u) \times D(u)$ does not vanish. If the tangent plane to the surface at any point is tangent along the entire generator through the point, then the surface is called a *developable surface*. For example, a cylinder is a developable surface, as is a cone. A hyperboloid of one sheet is an example of a ruled surface which is not a developable surface.

diagonal For any set X, the subset $\Delta = \{(x, x) : x \in X\}$ of $X \times X$. This is also commonly referred to as the *relation of equality* on X. *See* relation.

diagonal intersection If κ is a regular uncountable cardinal and $(S_\alpha, \ \alpha < \kappa)$ is a sequence of subsets of κ, the *diagonal intersection* of this sequence, denoted $\Delta\{S_\alpha : \alpha < \kappa\}$, is $\{\beta : \beta \in \bigcap_{\alpha < \beta} S_\alpha\}$.

diagram A *diagram*

$$\overset{a}{\cdot} \overset{b}{\to} \overset{}{\cdot} \to \ \cdots$$

consists of vertices $a, b, \ldots$ and arrows. The vertices represent sets $A, B, \ldots$ and the arrows represent functions between the sets.

diagram in the category In the category $\mathcal{C}$, a diagram in which the vertices represent objects of $\mathcal{C}$ and the arrows represent morphisms of $\mathcal{C}$. *See* diagram, category.

diameter The greatest distance between any two points of the body in question.

diametral plane A plane containing all midpoints of a set of parallel chords of a surface.

diamond A strengthening of the Continuum Hypothesis (denoted $\diamondsuit$) which asserts that there is a sequence of sets $S_\alpha \subseteq \alpha$ for $\alpha < \omega_1$, called the *diamond* sequence, which captures all subsets of ω_1 in a certain way. Given any $X \subseteq \omega_1$, the set of α where $X \cap \alpha = S_\alpha$ is stationary in ω_1. In other words, there are a large number of α where S_α is the same as X up to α.

To see why this implies the Continuum Hypothesis, notice that if $X \subseteq \omega$, then

$$X \cap \alpha = X$$

for any $\alpha \geq \omega$. But $C = [\omega, \omega_1)$ is closed and unbounded, so there is an $\alpha \in C \cap S$ where $S_\alpha = X \cap \alpha = X$. That is, each subset of the natural numbers appears in the diamond sequence, so the number of subsets is at most ω_1.

Other consequences of $\diamondsuit$ include the negation of Suslin's Hypothesis; i.e., $\diamondsuit$ implies there is an ω_1-Suslin tree.

difference of sets For two sets X and Y, the set $X \backslash Y$ consisting of all elements of X which are not elements of Y. More precisely,

$$X \backslash Y = \{x \in X : x \notin Y\}.$$

differentiable structure A compatible way of assigning, to each point in a space, a homeomorphism from a neighborhood of that point to an open subset of n-dimensional real space $\mathbf{R}^n$, or of n-dimensional complex space $\mathbf{C}^n$.

differential geometry The body of geometry that investigates curves and surfaces in the immediate neighborhood of one of their points, using calculus, and analyzes what is implied about the curve or surface as a whole on the basis of this local behavior. A more advanced aspect of *differential geometry* is the possibility of constructing geometrical systems determined solely by concepts and postulates that affect only the immediate neighborhood of each point of the system.

differential invariant An expression that consists of certain functions, partial derivatives, and differentials, which is invariant with respect to certain transformations.

dihedral angle   The angle between two planes, measured as the angle in a plane perpendicular to the line of intersection of the two planes. If the planes do not intersect, or coincide, the *dihedral angle* is zero.

dilatation   An affinity that possesses a fixed point and maps every line onto a parallel of itself.

dimension   Any one of many possible different topologically invariant measures of the size of a topological space. Different definitions of *dimension* include the Lebesgue dimension, the homological dimension, the cohomological dimension, and the large and small inductive dimensions. The large inductive dimension, which agrees with the Lebesgue and small inductive dimensions when the space is separable and metrizable, is defined inductively as follows. We say that the empty set has dimension -1. Assuming that we have defined all spaces of dimension $\leq n$, we say that a space X has dimension $\leq n + 1$ if, for any disjoint closed subsets C and D of X, there is a closed subset T of X with dimension $T \leq n$ such that $X \backslash T$ is the union of two disjoint open subsets, one containing C and the other containing D. We then say that a topological space has dimension n if it has dimension $\leq n$ but it does not have dimension $\leq n - 1$.

dimension function   A function $d : L \rightarrow \mathbf{Z}$ from a lattice L to the nonnegative integers satisfying the conditions (i.) $d(x+y)+d(xy) = d(x)+d(y)$ for all $x, y \in L$, and (ii.) if $[x, y]$ is an elementary interval in L, then $d(y) = d(x)+1$.

dimension of a complex   Let X be a CW-complex and let E be the set of cells of X. The *dimension* of X is given by:

$$\dim X = \sup\{\dim(e) : e \in E\}.$$

We say that X is *finite dimensional* if $\dim X$ is finite and *infinite dimensional* otherwise.

dimensions of a rectangle   The dimensions that fully describe a rectangle, namely length and width.

Dimension Theorem of Affine Geometry
Given affine space $\mathbf{A}^n$, with nondisjoint subspaces $\mathbf{A}^r$ and $\mathbf{A}^s$ of dimension r and s, respectively, then $r + s = \dim(\mathbf{A}^r \cup \mathbf{A}^s) + \dim(\mathbf{A}^r \cap \mathbf{A}^s)$.

dimension theory   The branch of topology devoted to the definition and study of the notion of dimension in various classes of topological spaces.

dimension type   Two topological spaces X and Y have the same *dimension type* if X is homeomorphic to a subspace of Y and Y is homeomorphic to a subspace of X.

dimension zero   A topological space X has *dimension zero* if it has the discrete topology. *See* topological dimension.

Dini surface   A helicoidal surface in three-dimensional Euclidean space, which is the surface of revolution of a tractrix.

directed set   A set D with a partial order $\preceq$, such that for all $a, b \in D$ there exists an element $c \in D$ so that $a \preceq c$ and $b \preceq c$.

directrix   The polar, with respect to the reciprocating circle of a conic section, of the center of the reciprocal circle of the conic section; this applies when the conic section is regarded as the reciprocal of a circle. Alternatively, if the conic section is regarded as a curve generated by a point moving in the plane such that the ratio of its distance from a fixed point to a fixed line remains constant, then the *directrix* of the conic section is that fixed line.

directrix of Wilczynski   Two straight lines associated with a normal frame in projective differential geometry.

Dirichlet convolution   The arithmetic function $f * g$ defined by $(f * g)(n) = \sum_d f(d)g(\frac{n}{d})$, where f and g are arithmetic functions, and d ranges over the divisors of n. (*See* arithmetic function.) For example, if $f = \phi$, the Euler phi function, and $g = \tau$, the number of divisors function,

$$(\phi * \tau)(10)$$

$$= \phi(1)\tau(10) + \phi(2)\tau(5) + \phi(5)\tau(2) + \phi(10)\tau(1)$$

$$= 18 .$$

In fact, $\phi * \tau = \sigma$, the sum of divisors function. The *Dirichlet convolution* is also called the Dirichlet product.

Dirichlet inverse   The *Dirichlet inverse* of an arithmetic function f is a function f^{-1} such that the Dirichlet convolution $f * f^{-1} = I$, the identity function. (*See* arithmetic function, Dirichlet convolution.) A function f has a Dirichlet inverse if and only if $f(1) \neq 0$. When it exists, the inverse is unique. For example, μ, the Möbius function, and u, the unit function, are Dirichlet inverses of one another.

Dirichlet multiplication   The operation under which the Dirichlet convolution of two arithmetic functions is computed. It is commutative and associative. In fact, the set of arithmetic functions f such that $f(1) \neq 0$ forms a group under this operation. *See also* Dirichlet convolution.

Dirichlet product   *See* Dirichlet convolution.

discrete linear ordering   A linear ordering $\leq$ on a set A such that

(i.) every element $x \in A$ that has a successor (i.e., an element $y \in A$ such that $x < y$) has an immediate successor (i.e., there exists $z \in A$ such that z is a successor of x and there does not exist $y \in A$ with $x < y < z$), and

(ii.) every element $x \in A$ that has a predecessor (i.e., an element $y \in A$ such that $y < x$) has an immediate predecessor (i.e., there exists $z \in A$ such that z is a predecessor of x and there does not exist $y \in A$ with $z < y < x$).

The usual ordering $\leq$ on the set $\mathbf{N}$ of natural numbers is discrete.

discrete topology   The topology on a set X, consisting of all subsets of X. That is, every subset is open in X.

disjoint sets   Two sets X and Y which have no common elements. Symbolically, X and Y are disjoint if

$$X \cap Y = \emptyset .$$

disjunctive normal form   A propositional (sentential) formula of the form

$$\bigvee_{i=1}^{n} \left(\bigwedge_{j=1}^{m_i} A_{ij} \right) ;$$

i.e.,

$$(A_{11} \wedge \cdots \wedge A_{1m_1}) \vee \cdots \vee$$
$$(A_{n1} \wedge \cdots \wedge A_{nm_n}) ,$$

where each A_{ij}, $1 \leq i \leq n$, $1 \leq j \leq m_i$, is either a sentence symbol or the negation of a sentence symbol. Every well-formed propositional formula is logically equivalent to one in *disjunctive normal form*.

For example, if A, B, and C are sentence symbols, then the well-formed formula $((A \rightarrow B) \rightarrow C)$ is logically equivalent to $(A \wedge B \wedge C) \vee (A \wedge \neg B \wedge C) \vee (A \wedge \neg B \wedge \neg C) \vee (\neg A \wedge B \wedge C) \vee (\neg A \wedge \neg B \wedge C)$, which is a formula in disjunctive normal form.

distance   Part of the definition of a metric space M. The distance function on M, $d : M \times M \rightarrow \mathbf{R}$, must be nonnegative-valued and satisfy (i.) $d(P_1, P_2) = 0$ if and only if $P_1 = P_2$, (ii.) $d(P_1, P_2) = d(P_2, P_1)$, and (iii.) $d(P_1, P_2) + d(P_2, P_3) \geq d(P_1, P_3)$ for all $P_1, P_2, P_3 \in M$. Then the distance between two points P_1 and P_2 is $d(P_1, P_2)$. We may then also define the distance between any two subsets S and T of M to be the greatest lower bound of the set $\{d(s, t) : s \in S, t \in T\}$.

In three-dimensional Euclidean space, the distance function is given by

$$d((x_1, y_1, z_1), (x_2, y_2, z_2))$$

$$= \sqrt{(x_2 - x_1)^2 + (y_2 - y_1)^2 + (z_2 - z_1)^2} .$$

distance function   A function $d : X \times X \rightarrow \mathbf{R}$, where X is a topological space and $\mathbf{R}$ is the real numbers, which satisfies the following three conditions:

(i.) $d(x, y) \geq 0$, and $d(x, y) = 0$ if and only if $x = y$;

(ii.) $d(x, y) = d(y, x)$; and

(iii.) $d(x, y) + d(y, z) \geq d(x, z)$.

This last condition, known as the triangle inequality, generalizes the principle of plane geometry that the length of any side of a triangle is not longer than the sum of the lengths of the other two sides.

division algorithm   If a and $b \neq 0$ are in $\mathbf{Z}$, there exist unique integers r so that $a = bq + r$ and $0 \leq r < b$. A clever repeated application of the *division algorithm,* known as Euclid's algorithm, leads to the computation of the greatest common divisor of the integers a and b. *See* Euclidean algorithm.

Division algorithms hold in other rings, such as the polynomials with real coefficients.

divisor   If a and b are elements of a ring and there exists an element c in that ring satisfying $bc = a$, then b (similarly, c) is a *divisor* of a. For example, in the ring of integers, 6 is a divisor of 24 since $6 \times 4 = 24$ (and 4, 6, and 24 are all integers), while 5 is not a divisor of 24 since there is no integer c so that $5c = 24$. However, in the ring $\mathbf{Z}_{36}$ (the integers mod 36), 5 *is* a divisor of 24 since $5 \times 12 = 24$ in $\mathbf{Z}_{36}$ (alternatively, $5 \times 12 \equiv 24 \pmod{36}$).

dodecagon   A polygon having 12 sides.

dodecahedron   A polyhedron with 12 faces. The *regular dodecahedron,* the regular convex polyhedron having 12 pentagonal faces, 30 edges, and 20 vertices, is one of the five platonic solids.

domain   For a binary relation R on two sets X and Y, the set

$$\mathrm{dom}(R) = \{x : (x, y) \in R \text{ for some } y \in Y\}.$$

Commonly, the relation R is a function f, $R = \{(x, y) : y = f(x)\}$, and the domain of the function f is the domain of the relation R, in this case. *See* relation.

double angle formulas   The trigonometric identities $\sin 2\theta = 2 \sin \theta \cos \theta$ and $\cos 2\theta = 1 - 2 \sin^2 \theta$.

dual   The *dual* of a concept represented by a diagram is the diagram in which the vertices are the same, but all arrows are reversed. *See* diagram.

dual bundle   Given a bundle ξ, with projection

$$\pi : E \to B ,$$

the *dual bundle* ξ^* of ξ has the projection

$$\pi' : E' \to B ,$$

with $E' = \cup_{p \in B}[\pi^{-1}(p)]^*$, where $[\pi^{-1}(p)]^*$ denotes the dual space of $\pi^{-1}(p)$, and π' takes each $[\pi^{-1}(p)]^*$ to p.

dual category   The dual of a category $\mathcal{C}$ (also known as the opposite of $\mathcal{C}$) is the category $\mathcal{C}^{op}$ which satisfies the following properties:

(i.) $\mathrm{Obj}(\mathcal{C}^{op}) = \mathrm{Obj}(\mathcal{C})$;

(ii.) $\mathrm{Hom}_{\mathcal{C}^{op}}(A, B) = \mathrm{Hom}_{\mathcal{C}}(B, A)$.

Composition of morphisms in $\mathcal{C}^{op}$ is defined by $g^{op} f^{op} = (fg)^{op}$. *See* category.

dual complex   The set of dual cells of simplices of a complex. More specifically, consider a simplicial complex C. Let C' be the barycentric subdivision of C and, for any q-simplex σ of C, let $C(\sigma)$ denote the union of all $(n - q)$-simplices (n being the dimension of the manifold) of C'. Then the set $C^* = \{C(\sigma) : \sigma \in C\}$ is the *dual complex* of C.

dual convex cone   Given a convex cone $C \subseteq \mathbf{R}^n$, the set $\{x \in \mathbf{R}^n : (x, y) \leq 0 \text{ for all } y \in C\}$. Here (x, y) denotes the inner product of x and y.

Dupin indicatrix   If M is a surface in $\mathbf{R}^3$, and P is a point in M, then a plane parallel to the tangent plane to M at P and very close to the tangent plane will intersect M in a curve that is approximately a quadratic curve. The *Dupin indicatrix* is a quadratic curve that is similar to this curve of intersection. If the principal curvatures κ_1 and κ_2 of the surface at P are both positive, then the Dupin indicatrix is given by the ellipse $\kappa_1 x^2 + \kappa_2 y^2 = 1$. If $\kappa_1 > 0 > \kappa_2$, then the Dupin indicatrix is the pair of hyperbolas $\kappa_1 x^2 + \kappa_2 y^2 = \pm 1$. When one of the principal curvatures is 0, the indicatrix is a pair of parallel straight lines.

duplication of cube   One of the "Three Famous Greek Problems" from the classical Greek geometers. In this problem, a cube is to be constructed with double the volume of a given cube. It can be proved that this construction is impossible using a straight edge and compass alone.

dyadic compactum   Let X be the discrete space with two points. The infinite product of copies of X, with the product topology, is the *dyadic compactum*. It is a compact, uncountable, totally disconnected, Hausdorff space, homeomorphic to the Cantor set. *See also* Cantor set.

E

eccentric angle For an ellipse, the angle θ, where the ellipse is described parametrically by the equations $x = a\cos\theta$, $y = b\sin\theta$. Similarly, the *eccentric angle* at (x, y) for a hyperbola described parametrically by the equations $x = a\sec\phi$, $y = b\tan\phi$ is ϕ.

eccentric circles (1) For an ellipse, the circles centered at the center of the ellipse with diameters equal to the lengths of the major and minor axes of the ellipse.

(2) The two *eccentric circles* of a hyperbola are those with center at the origin of the hyperbola, and with diameters equal to the lengths of the transverse and conjugate axes of the hyperbola.

eccentricity For a conic section, the ratio $\frac{OA}{r}$, when the conic section in question is regarded as the reciprocal of a circle with radius r and center A with respect to the circle having center O. Alternatively, if the conic section is regarded as a curve generated by a point moving in the plane such that the ratio of its distance from a fixed point to a fixed line remains constant, then the *eccentricity* of the conic is that distance ratio.

effective Informally, the term *effective* is often used as in the definition of effective procedure, as a synonym for "algorithmic". Formally, the term effective is synonymous with computable, or recursive. *See* effective procedure, computable, recursive.

effectively enumerable A set A of natural numbers such that there is an effective procedure which, when given a natural number n, will output 1 after finitely many steps if $n \in A$ and will run forever otherwise. Alternatively, A is *effectively enumerable* if there is an effective procedure that lists the elements of A. In other words, for an effectively enumerable set A, if $n \in A$,

one will find out eventually, but if $n \notin A$, then one will never know for sure.

For example, the set of all natural numbers n such that there exists a consecutive run of exactly n 5s in the decimal expansion of π is effectively enumerable.

This notion is intuitive; the corresponding formal notion is recursively enumerable, also known as computably enumerable. *See* recursively enumerable.

effective procedure An *effective procedure* (or algorithm) is a finite, precisely given list of instructions which is deterministic; i.e., at any step during the execution of the instructions, there must be at most one instruction that can be applied. This notion is intuitive for a corresponding formal mathematical notion. *See* computable, recursive.

Eilenberg-Steenrod Axioms Let $\mathcal{T}$ be a category of pairs of topological spaces and continuous maps and let $\mathcal{A}$ denote the category of Abelian groups. Suppose we have the following:

(i.) A functor $H_p : \mathcal{T} \to \mathcal{A}$ for each integer $p \geq 0$, whose value is denoted $H_p(X, A)$. If $f : (X, A) \to (Y, B)$ is a continuous map, let $(f_*)_p$ denote the induced map from $H_p(X, A)$ to $H_p(Y, B)$.

(ii.) A natural transformation

$$\partial_p : H_p(X, A) \to H_{p-1}(A)$$

for each integer $p \geq 0$, where A denotes the pair $(A, \emptyset)$.

These functors and natural transformations must satisfy the following three axioms from category theory. All pairs are in $\mathcal{T}$.

Axiom 1. If i is the identity, then $(i_*)_p$ is the identity for each p.

Axiom 2. $((k \circ h)_*)_p = (k_*)_p \circ (h_*)_p$.

Axiom 3. If $f : (X, A) \to (Y, B)$, then the following diagram is commutative:

$$
\begin{array}{ccc}
H_p(X, A) & \xrightarrow{(f_*)_p} & H_p(Y, B) \\
\partial p \downarrow & & \partial p \downarrow \\
H_{p-1}(A) & \xrightarrow{((f|_A)_*)_p} & H_{p-1}(B) \,.
\end{array}
$$

The *Eilenberg-Steenrod axioms* are the following five axioms:

Exactness Axiom. The sequence

$$\cdots \to H_p(A) \xrightarrow{(i_*)_p} H_p(X) \xrightarrow{(\pi_*)_p} H_p(X, A)$$

$$\xrightarrow{\partial_p} H_{p-1}(A) \rightarrow \dots$$

is exact, where $i : A \rightarrow X$ and $\pi : X \rightarrow (X, A)$ are the inclusion maps.

Homotopy Axiom. If h and k are homotopic, then $(h_*)_p = (k_*)_p$ for each p.

Excision Axiom. Given (X, A), let U be an open subset of X such that $\bar{U} \subset \text{Int} A$. If $(X \setminus U, A \setminus U)$ is in $\mathcal{A}$, then the inclusion $(X \setminus U, A \setminus U) \hookrightarrow (X, A)$ induces an isomorphism

$$H_p(X \setminus U, A \setminus U) \simeq H_p(X, A) .$$

Dimension Axiom. If P is a one-point space, the $H_p(P) = \{0\}$ for $p \neq 0$ and $H_0(P) \simeq \mathbf{Z}$.

Axiom of Compact Support. If $z \in H_p(X, A)$, there is a pair (X_0, A_0) in $\mathcal{T}$ with X_0 and A_0 compact, such that z is in the image of the homomorphism $H_p(X_0, A_0) \rightarrow H_p(X, a)$ induced by the inclusion $(X_0, A_0) \hookrightarrow (X, A)$.

Any theory that satisfies these axioms is called a *homology theory* on $\mathcal{T}$. (*See* homology theory.) The first homology theory was defined for the category of compact polyhedra. Later several other homology theories, such as *singular homology*, were defined. Eilenberg and Steenrod then showed that the above axioms completely classified the homology groups on the class of polyhedra. There are also similar axioms for *cohomology theory*, except for the Axiom of Compact Support.

elementarily equivalent structures Two structures A and B in the language L such that, for every sentence ϕ of L, $A \models \phi$ if and only if $B \models \phi$; that is, ϕ is true in A if and only if it is true in B. Elementary equivalence (written $A \equiv B$) expresses the property that L cannot distinguish between the structures A and B.

elementary diagram The theory of all sentences which hold in a model A, using an extra constant symbol for each element of A. More precisely, let A be a model in the language L, and let L_A be the expansion of L which adds a new constant symbol c_a for each $a \in A$. Then the *elementary diagram* of A is the set of all L_A-sentences which are true in the model A with each c_a interpreted by a.

elementary embedding Let $\mathcal{L}$ be a first order language, and let $\mathcal{A}$ and $\mathcal{B}$ be structures

for $\mathcal{L}$, where A and B are the universes of $\mathcal{A}$ and $\mathcal{B}$, respectively. An *elementary embedding* of $\mathcal{A}$ into $\mathcal{B}$ is an embedding h of $\mathcal{A}$ into $\mathcal{B}$ with the property that for every well-formed formula φ with free variables $v_1, \dots, v_n$ and every n-tuple $a_1, \dots, a_n$ of elements of A, if $\models_{\mathcal{A}} \varphi[a_1, \dots a_n]$, then $\models_{\mathcal{B}} \varphi[h(a_1), \dots, h(a_n)]$. *See* embedding, satisfy (for the definition of the notation used here).

elementary substructure Let $\mathcal{L}$ be a first order language, and let $\mathcal{A}$ and $\mathcal{B}$ be structures for $\mathcal{L}$, where A is the universe of $\mathcal{A}$. The structure $\mathcal{A}$ is an *elementary substructure* of $\mathcal{B}$ if
   (i.) $\mathcal{A}$ is a substructure of $\mathcal{B}$, and
   (ii.) for all well-formed formulas φ with free variables from among $v_1, \dots, v_n$ and all n-tuples $a_1, \dots, a_n$ of elements of A, if $\models_{\mathcal{A}} \varphi[a_1, \dots, a_n]$, then $\models_{\mathcal{B}} \varphi[a_1, \dots, a_n]$.
   See satisfy (for the definition of the notation used here).

If $\mathcal{A}$ is an elementary substructure of $\mathcal{B}$, then $\mathcal{B}$ is an elementary extension of $\mathcal{A}$. The term elementary submodel is sometimes synonymous with elementary substructure.

As an example, let $\mathcal{L}$ be the first order language with equality whose only predicate symbol is $<$. Let $\mathcal{Q}$ be the structure whose universe is the set $\mathbf{Q}$ of rational numbers and where $<$ is interpreted in the usual way, and let $\mathcal{R}$ be the structure whose universe is the set $\mathbf{R}$ of real numbers and where $<$ is interpreted in the usual way. Then $\mathcal{Q}$ is an elementary substructure of $\mathcal{R}$.

element of a set One of the objects x that makes up the set X, written $x \in X$. *See* set.

element of cone Any line that lies on the surface of a given cone and contains its vertex.

element of cylinder The generator of a given cylinder in any fixed position, where the cylinder is thought of as being generated by a straight line moving along a given curve while remaining parallel to a fixed line.

ellipse A proper conic section formed by the intersection of a plane with one nappe of the cone. Alternatively, a conic section with eccentricity less than one.

ellipsoid   A surface whose intersection with any plane is either a point, a circle, or an ellipse.

elliptic cone   The set of points consisting of all the lines passing through a fixed ellipse and a fixed point not in the plane of the ellipse.

elliptic cylinder   The set of points consisting of all the lines passing through a fixed ellipse and parallel to a fixed line not parallel to the plane of the ellipse.

elliptic point   A point on a surface at which the centers of curvature are all on the same side of the surface normal; the normal sections are all concave or all convex.

elliptic surface   Any type of Riemann surface that can be mapped conformally on the closed complex plane. More generally, a nonsingular surface E having a surjective morphism

$$\pi : E \to S$$

onto a nonsingular curve S whose generic fiber is a nonsingular elliptic curve.

elliptic transformation   A linear fractional transformation

$$z \mapsto \frac{az + b}{cz + d}$$

on the complex numbers $\mathbf{C}$, where $a + d$ is real, and discriminant $(a + d)^2 - 4$ is negative.

embedding   (1) Let $\mathcal{L}$ be a first order language, and let $\mathcal{A}$ and $\mathcal{B}$ be structures for $\mathcal{L}$, with universes A and B for $\mathcal{A}$ and $\mathcal{B}$, respectively. A function $h : A \to B$ is an *embedding* if h is injective and

(i.) for each n-ary predicate symbol P and every $a_1, \ldots a_n \in A$,

$$(a_1, \ldots, a_n) \in P^{\mathcal{A}}$$
$$\Leftrightarrow (h(a_1), \ldots, h(a_n)) \in P^{\mathcal{B}} ,$$

(ii.) for each constant symbol c,

$$h(c^{\mathcal{A}}) = c^{\mathcal{B}},$$

and

(iii.) for each n-ary function symbol f and every $a_1 \ldots, a_n \in A$,

$$h(f^{\mathcal{A}}(a_1, \ldots, a_n)) = f^{\mathcal{B}}(h(a_1), \ldots, h(a_n)).$$

If there is an embedding of $\mathcal{A}$ into $\mathcal{B}$, then $\mathcal{A}$ is isomorphic to a substructure of $\mathcal{B}$.

(2) An injective map f of a space X into a space Y such that if $Z = f(X)$, then the map $f' : X \to Z$, obtained by restricting the codomain of f, is a homeomorphism.

empty set   A set denoted $\emptyset$ which has no elements.

enumeration   An *enumeration of a set A* is a surjection $f : \mathbf{N} \to A$; i.e., a function f which has domain $\mathbf{N}$ and range A. Such a function is called an enumeration because it "lists" the elements of A. An enumeration need not be an injection (i.e., one-to-one), nor list the elements of A in any particular order.

equal geometric figures   Two figures such that one can be moved coincident with the other via a transformation.

equal sets   Two sets A and B which have the same elements; that is, if for all x, $x \in A$ if and only if $x \in B$. In formal ZF (Zermelo-Fraenkel) set theory, this is called the Axiom of Extensionality.

equiangular polygon   A polygon whose interior angles all have the same measure.

equiangular spiral   A spiral given by the polar equation $r = e^{k\theta}$, where k is a constant. Also known as the logarithmic or exponential spiral, or the spiral of Bernoulli.

equidistant   A set of objects such that any pair of objects in the set is the same distance apart as any other pair of objects in the set.

equilateral   A figure with sides, all of which have the same length.

equilateral triangle   A triangle having all three sides congruent.

equinumerous sets   Two sets A and B such that there is a bijection, or one-to-one correspondence, between them; i.e., there is a function $f : A \to B$ such that f is both injective and surjective. For example, if $\mathbf{N}$ denotes the set of natural numbers, $\mathbf{Q}$ denotes the set of rational numbers, and $\mathbf{R}$ denotes the set of real numbers, then $\mathbf{N}$ and $\mathbf{Q}$ are equinumerous, while $\mathbf{Q}$ and $\mathbf{R}$ are not equinumerous.

equipollent sets   Two sets A and B which have a bijection $f : A \to B$ between them.

equivalent bases for a topological space   Let X be a topological space. The bases $\mathcal{B}$ and $\mathcal{B}'$ are *equivalent* if they generate the same topology on X. That is, for all $B \in \mathcal{B}$, if $x \in B$ then there exists $B' \in \mathcal{B}'$ so that $x \in B' \subset B$. Conversely for all $B' \in \mathcal{B}'$, if $x \in B'$ then there exists $B \in \mathcal{B}$ so that $x \in B \subset B'$.

equivalent sets   Two sets A and B such that there exists a bijection $f : A \to B$.

Erlangen Program   The name given to a method for studying the geometry of a space X.

Initiated by Felix Klein, the program proposed a study of the geometric properties of a space X that remain invariant under a specified group of one-to-one continuous transformations of the space.

For example, the geometry of the Euclidean plane can be described by the group of rigid motions of $\mathbf{R}^2$ that take congruent figures to one another.

escribed circle of a triangle   A circle tangent to one side of the triangle as well as to the extensions of the other two sides.

Euclidean   Satisfying the postulates of Euclid's Elements.

Euclidean algorithm   A method for determining the greatest common divisor of two nonzero integers using repeated application of the division algorithm. *See* division algorithm.

To find $\gcd(10, 46)$, begin by using the division algorithm to determine the remainder obtained when 44 is divided by 12:

$$46 = 4(10) + 6 .$$

Next, repeat the division algorithm with 10 and 6 (the dividend and remainder from above):

$$10 = 1(6) + 4 .$$

This procedure (repeating the division algorithm with the previous dividend and remainder) is repeated until a 0 remainder is obtained (note that this is guaranteed to occur eventually, since the remainders are necessarily decreasing). To continue the illustration, repeat the division algorithm with 6 and 4 to obtain,

$$6 = 1(4) + 2 ,$$

then apply the division algorithm one more time to get

$$4 = 2(2) + 0 .$$

The last nonzero remainder will always be the greatest common divisor of the two original integers.

To illustrate the algorithm more succinctly,

$$\begin{aligned}
46 &= 4(10) + 6 \\
10 &= 1(6) + 4 \\
6 &= 1(4) + 2 \\
4 &= 2(2) + 0
\end{aligned}$$

As 2 is the last nonzero remainder, we conclude that $\gcd(10, 46) = 2$.

Also known as Euclid's algorithm.

Euclidean geometry   Ordinary plane or three-dimensional geometry. More generally, it can refer to any geometry in which each point is uniquely described by an ordered set of n numbers, the coordinates of the point, and where the distance $d(x, y)$ between two points $x = (x_1, \ldots, x_n)$ and $y = (y_1, \ldots, y_n)$ is given by $d(x, y) = \sqrt{\sum_{i=1}^{n}(y_i - x_i)^2}$.

Euclidean plane   Two-dimensional Euclidean space, in which each point is uniquely described by an ordered pair of real numbers (x, y), and distance between points $P_1 = (x_1, y_1)$ and $P_2 = (x_2, y_2)$ is given by

$$d(P_1, P_2) = \sqrt{(x_2 - x_1)^2 + (y_2 - y_1)^2} .$$

Euclidean polyhedron   In $\mathbf{R}^3$, a solid bounded by polygons. More generally, the set of points belonging to the simplices of a Euclidean simplicial complex in $\mathbf{R}^n$.

Euclidean space   A space that has a Euclidean geometry. *See* Euclidean geometry.

Euler characteristic   Let K be a simplicial complex of dimension n and let α_m be the number of simplices of dimension m. Then the *Euler-Poincaré characteristic*, $\chi(K)$, of K is defined by:

$$\chi(K) = \sum_{m=0}^{n} (-1)^m \alpha_m .$$

The most common version of the Euler-Poincaré characteristic occurs in the case where K has dimension two. If we let V be the number of vertices, E be the number of edges, and F be the number of faces of K, then $\chi(K) = V - E + F$.

The Euler-Poincaré characteristic is an invariant of the complex; that is, it is independent of the triangulation of the complex K.

If β_p is the pth *Betti number* of K, that is $\beta_p = \text{rank} H_p(K)/T_p(K)$ where $T_p(K)$ is the torsion subgroup of $H_p(K)$, then

$$\chi(K) = \sum_{p=0}^{n} (-1)^p \beta_p .$$

Euler phi function   The arithmetic function, denoted φ, which, for any positive integer n, returns the number of positive integers less than or equal to n which are relatively prime to n. (*See* arithmetic function.) That is, $\phi(n) = \#\{i : 1 \le i \le n \text{ and } (i, n) = 1\}$. For example, $\phi(6) = 2, \phi(13) = 12$. The value of $\phi(n)$ is even for all $n > 1$. It is multiplicative; its value at a prime power is given by $\phi(p^i) = p^{i-1}(p - 1)$. It is also called the totient function.

Euler-Poincaré class   Given an orientable vector bundle ξ, with base space B, on $\mathbf{R}^n$, the primary obstruction in $H^n(B; \mathbf{Z})$ for constructing a cross-section of the associated $(n-1)$-sphere bundle. The *Euler-Poincaré* class of a manifold is that of its tangent bundle.

Euler-Poincaré formula   *See* Euler characteristic.

Euler product   If f is a multiplicative function (a real or complex valued function defined on the positive integers with the property that if $\gcd(m, n) = 1$, then $f(mn) = f(m)f(n)$) and the series $\sum_{n=1}^{\infty} f(n)$ converges absolutely, then

$$\sum_{n=1}^{\infty} f(n) = \prod_{p} \left(1 + f(p) + f(p^2) + \cdots \right) ,$$

where the product is taken over all primes. This product is called the *Euler product* of the series.

If f is *completely* multiplicative ($f(mn) = f(m)f(n)$ for all m, n), in which case $f(p^k) = f(p)^k$ for each k and p, then the Euler product above can be simplified using our knowledge of geometric series and we have

$$\sum_{n=1}^{\infty} f(n) = \prod_{p} \frac{1}{1 - f(p)} .$$

Euler product formula   The Euler product for certain Dirichlet series. *See* Euler product.

For example, using the Euler product (with $f(n) = 1$ for all n), we can express the Riemann zeta function as a product. Namely,

$$\zeta(s) = \prod_{p} \left(1 - \frac{1}{p^s} \right)^{-1}$$

for all real numbers $s > 1$.

Euler's criterion   Let p be an odd prime. If p is not a divisor of the integer a, then a is a quadratic residue of p if and only if $a^{\frac{p-1}{2}}$ is one more than a multiple of p (note that $a^{\frac{p-1}{2}}$ is always either one more or one less than a multiple of p by Fermat's Little Theorem).

Euler's summation formula   A formula that specifies the error involved when a partial sum of an arithmetic function is approximated by an integral. Specifically, the formula states that if a and b are real numbers with $a < b$ and f is continuously differentiable on the interval

$[a, b]$, then

$$\sum_{a < k \leq b} f(k) = \int_a^b f(x)\, dx$$
$$+ \int_a^b f'(x)(x - [x])\, dx$$
$$+ (b - [b]) f(b)$$
$$- (a - [a]) f(a).$$

Here, $[x]$ denotes the greatest integer less than or equal to x (the so-called greatest integer function).

The Euler-Maclaurin summation formula is a special case of Euler's formula when a and b are integers. Namely,

$$\sum_{k=a}^b f(k) = \int_a^b f(x)\, dx$$
$$+ \int_a^b f'(x)(x - [x] - \frac{1}{2})\, dx$$
$$+ \frac{1}{2} f(a) - \frac{1}{2} f(b).$$

Euler's Theorem of Polyhedrons   For Euclidean space, this theorem states that $V - E + F = 2$ for any simple polyhedron, where $V =$ number of vertices, $E =$ number of edges, and $F =$ number of faces in the polyhedron. This theorem may be generalized to state that, for any finite CW complex, $\alpha_0 - \alpha_1 + \alpha_2 = 2$, where $\alpha_i =$ number of i-cells of the CW complex.

exact functor   A diagram

$$0 \longrightarrow A \overset{f}{\longrightarrow} B \overset{g}{\longrightarrow} C \longrightarrow 0$$

in the category of modules is exact if f is injective, g is surjective, and the kernel of g is equal to the image of f. An *exact functor* is an additive functor $F : \mathcal{C} \to \mathcal{D}$ between categories of modules satisfying the property that the exactness of the diagram

$$0 \longrightarrow A \overset{f}{\longrightarrow} B \overset{g}{\longrightarrow} C \longrightarrow 0$$

implies the exactness of either

$$0 \longrightarrow F(A) \overset{F(f)}{\longrightarrow} F(B) \overset{F(g)}{\longrightarrow} F(C) \longrightarrow 0$$

or

$$0 \longrightarrow F(C) \overset{F(g)}{\longrightarrow} F(B) \overset{F(f)}{\longrightarrow} F(A) \longrightarrow 0,$$

depending on whether F is covariant or contravariant, respectively. *See* diagram, additive functor, covariant functor.

exact sequence of groups   A finite sequence of groups

$$A_0 \overset{f_0}{\to} A_1 \overset{f_1}{\to} A_2 \overset{f_3}{\to} \cdots \overset{f_{n-1}}{\to} A_{n-1} \overset{f_n}{\to} A_n$$

is *exact* if $\mathrm{Im}(f_{i-1}) = \mathrm{Ker}(f_i)$ for $i = 1, 2, \ldots, n - 1$.

An infinite sequence of groups

$$\cdots \overset{f_{i-2}}{\to} A_{i-1} \overset{f_{i-1}}{\to} A_i \overset{f_i}{\to} A_{i+1} \overset{f_{i+1}}{\to} \cdots$$

is *exact* if $\mathrm{Im}(f_{i-1}) = \mathrm{Ker}(f_i)$ for all $i \in \mathbf{Z}$.

A *short exact sequence of groups* is an exact sequence $1 \to A \to B \to C \to 1$, where 1 denotes the trivial group, i is injective, and π is surjective. In this case we say that B is an *extension of A by C*.

existential quantifier   *See* quantifier.

existential sentence   Let $\mathcal{L}$ be a first order language and let σ be a sentence of $\mathcal{L}$. The sentence σ is an *existential sentence* if it has the form $\exists v_1 \ldots \exists v_n \alpha$, where α is quantifier-free, for some $n \geq 0$.

expansion of a language   Let $\mathcal{L}_1$ and $\mathcal{L}_2$ be first order languages. The language $\mathcal{L}_2$ is an *expansion* of $\mathcal{L}_1$ if $\mathcal{L}_1 \subseteq \mathcal{L}_2$; i.e., $\mathcal{L}_2$ has all the symbols of $\mathcal{L}_1$, together with additional predicate symbols, constant symbols, or function symbols.

Let $\mathcal{L}$ be a first order language, $\mathcal{A}$ a structure for $\mathcal{L}$, and let $X \subseteq A$, where A is the universe of $\mathcal{A}$. The expansion $\mathcal{L}_X$ is the expansion obtained from $\mathcal{L}$ by adding a new and distinct constant symbol c_a for each $a \in X$.

expansion of a structure   Let $\mathcal{L}_1$ and $\mathcal{L}_2$ be first order languages such that $\mathcal{L}_2$ is an expansion of $\mathcal{L}_1$, and let $\mathcal{A}$ be a structure for $\mathcal{L}_1$. An *expansion* of $\mathcal{A}$ to $\mathcal{L}_2$ gives interpretations in A, the universe of $\mathcal{A}$, of the additional predicate, constant, and function symbols in $\mathcal{L}_2$, while the interpretations in A of the symbols in $\mathcal{L}_1$ remain the same.

Let $\mathcal{L}$ be a first order language, $\mathcal{A}$ a structure for $\mathcal{L}$, and let $X \subseteq A$, where A is the universe of $\mathcal{A}$. The expansion $\mathcal{A}_X$ is the expansion of $\mathcal{A}$ to $\mathcal{L}_X$ by interpreting each new constant symbol c_a of $\mathcal{L}_X$, for each $a \in X$, by a; i.e., $c_a^{\mathcal{A}_X} = a$. This expansion is often denoted by $(\mathcal{A}, a)_{a \in X}$. *See also* expansion of a language.

extension   *See* substructure.

extension of a mapping   Suppose that $A \subset X$ and that $f : A \to Y$ is a map. Then $F : X \to Y$ is an *extension* of f if the restriction of F to A is equal to f; that is, $F(a) = f(a)$ for all $a \in A$.

extreme point   A point of a convex set in Euclidean space which is not the midpoint of a straight line joining two distinct points of the set.

F

F_σ set   A countable union of closed sets. *See* G_δ set.

face   A boundary polygon of a Euclidean polyhedron.   More generally, an $(n-1)$-dimensional subspace F of a convex cell C in n-dimensional affine space $\mathbf{A}^n$ such that F is the intersection of the boundary of C with an $(n-1)$-dimensional subspace of $\mathbf{A}^n$.

face angle   An angle between two edges of a polyhedron that share a vertex.

factor of integer   The integer b is a factor of the integer a if there exists an integer c so that $a = bc$. For example, 8 is a factor of 24 since $24 = 8 \times 3$, but 8 is not a factor of 36 since there is no integer c where $8c = 36$. *See also* divisor.

family   A *family of sets* is a function from an index set Λ to the set of subsets of a set X whose value at $\alpha \in \Lambda$ is denoted by X_α. Although the function can be denoted in the usual way as a set of ordered pairs $\{(\alpha, X_\alpha) : \alpha \in \Lambda\}$, it is completely specified by $\{X_\alpha : \alpha \in \Lambda\}$.

Farey arc   For a given positive integer n, construct the Farey series $\mathcal{F}_n$ of order n, that is, the ascending sequence of rational numbers $\frac{a}{b}$ between 0 and 1 with the property that $0 \le a \le b \le n$ and $\mathrm{GCD}(a, b) = 1$. Next, determine the *mediants* of all consecutive elements of $\mathcal{F}_n$. (*See* mediant.) A *Farey arc* is an interval of the form (p, q), where p and q are consecutive mediants. These intervals are usually visualized as arcs lying on the circle of circumference 1 on which the number x is represented by the point P_x lying x units counterclockwise from 0 (the "bottom" of the circle). Under the identification $0 = 1$ on this circle, we also include the arc $(\frac{n}{n+1}, \frac{1}{n+1})$ which runs from the "last" mediant to the first. Each of these arcs contains exactly one member of $\mathcal{F}_n$ The collection of Farey arcs

of order n is called the Farey *dissection* (of order n) of the circle. *See also* Farey dissection.

Farey dissection   The *Farey dissection* of order n is the collection of Farey arcs of order n. *See also* Farey arc.

Farey sequence   If n is a positive integer, the *Farey sequence* of order n (denoted by $\mathcal{F}_n$) is the sequence of rational numbers, listed in increasing order, whose denominator is no larger than n. For example,

$$\mathcal{F}_1 = (\ldots, \frac{-2}{1}, \frac{-1}{1}, \frac{0}{1}, \frac{1}{1}, \frac{2}{1}, \ldots)$$

$$\mathcal{F}_2 = (\ldots, \frac{-3}{2}, \frac{-1}{1}, \frac{0}{1}, \frac{1}{2}, \frac{1}{1}, \frac{3}{2}, \frac{2}{1}, \ldots)$$

$$\mathcal{F}_3 = (\ldots, \frac{-2}{3}, \frac{-1}{2}, \frac{-1}{3}, \frac{0}{1}, \frac{1}{3}, \frac{1}{2}, \frac{2}{3}, \frac{1}{1}, \frac{4}{3}, \ldots)$$

The Farey sequences can be used to make rational approximations.

Fermat number   A number of the form $F_n = 2^{2^n} + 1$, where n is a nonnegative integer. For example, $F_3 = 257$. The *Fermat number*s are named after Pierre de Fermat, a 17th century lawyer and amateur mathematician, who conjectured that all F_n are prime. The integers F_n, $n \le 4$, are in fact prime, but no other prime Fermat numbers are known. All F_n with $5 \le n \le 27$ (except $n = 24$) are known to be composite.

fiber   Any $f^{-1}(y)$ for $y \in Y$, where $f : X \to Y$ (that is, (X, Y, f)) is a fiber space. Sometimes spelled *fibre*. *See* fiber space.

fiber bundle   A *fiber bundle* (over a space X) is a fibration $f : F \to X$, namely a continuous surjective map such that X can be covered by open sets U_α over which the fibration is equivalent to the trivial one, the second projection of the Cartesian product $Y \times U_\alpha \to U_\alpha$, for Y a suitable space.

fiber product   In complete generality, the category-theoretical product $X \times_S Y$ of X with Y, where X, Y, and S are objects in a category $\mathcal{C}$, and X and Y (with given morphisms) are thought of as objects in the category $\mathcal{C}/S$. Here the category $\mathcal{C}/S$ has as its objects all morphisms to S. A morphism in $\mathcal{C}/S$ from $f : X \to S$ to

$g : Y \rightarrow S$ is any morphism h in C from X to Y such that $f = g \circ h$. Sometimes spelled *fibre product*.

fiber space   A triple consisting of two topological spaces X and Y, and a continuous map $f : X \rightarrow Y$, such that, for any cube $I^n = \{(x_1, \ldots x_n) : 0 \leq x_i \leq 1\}$, any mapping $\phi : I^n \rightarrow X$, and any homotopy $h_t : I^n \rightarrow Y$ with $f \circ \phi = h_0$, there is a homotopy $\phi_t : I^n \rightarrow X$ with $\phi_0 = \phi$ and $f \circ \phi_t = h_t$ for all t. Sometimes spelled *fibre space*.

fiber sum   Given two objects X and Y in a category C, the fiber product in the dual category C° of these two objects over another object S. Sometimes spelled *fibre sum*. *See* fiber product.

Fibonacci sequence   The recursive sequence $\{f_n\} = \{1, 1, 2, 3, 5, 8, \ldots\}$ defined by the initial conditions $f_0 = f_1 = 1$ and the recursive equation $f_{n+1} = f_n + f_{n-1}$. The mathematician Leonardo de Fibonacci originally developed this sequence to model the so-called "Rabbit Problem": Suppose that rabbits mature in one month, that the gestation period for rabbits is also one month, that a female rabbit always gives birth to a breeding pair of rabbits, and that rabbits never die. If a male and female rabbit are left on an uninhabited island at birth, how many pairs of rabbits will there be after a given number of months?

Initially, there is one pair of rabbits (we let f_0 denote the number (1) of pairs of immature rabbits at the end of the "0th" month). At the end of the first month, the rabbits have matured and are able to reproduce (after a gestation period of one month, the female will give birth) so that $f_1 = 1$, as well. At the end of the second month, the female rabbit has given birth to a pair of rabbits, so $f_2 = 2$. At the end of the third month, the new pair of rabbits has just matured and the first pair gives birth to another pair of rabbits. Thus, $f_3 = 3$. At the end of the fourth month, both the original pair and the first pair of offspring give birth to pairs of rabbits, so that $f_4 = 5$. It is left to the reader to show that the nth term of the *Fibonacci sequence* provides the number of pairs of rabbits which are alive at the end of the nth month.

filter   (1) A *filter on a set* S (or in $\mathcal{P}(S)$) is a collection $\mathcal{F}$ of subsets of S such that (i.) $S \in \mathcal{F}$, (ii.) $A, B \in \mathcal{F}$ implies $A \cap B \in \mathcal{F}$, for all A, B and (iii.) $A \in \mathcal{F}$ and $A \subseteq B$ implies $B \in \mathcal{F}$, for all A, B. A proper filter does not contain the empty set. For example, if S is any infinite set, the *Fréchet filter* on S is the set of all cofinite subsets of S. The Fréchet filter is a proper filter. Another important example of a proper filter is the *club filter* on some uncountable cardinal number κ. The club filter on κ is the set of all club subsets of κ.

(2) If $(\mathcal{B}, \vee, \wedge, \sim, 1, 0)$ is a Boolean algebra, $\mathcal{F} \subseteq \mathcal{B}$ is a *filter in* $\mathcal{B}$ if (i) $1 \in \mathcal{F}$, (ii) $a, b \in \mathcal{F}$ implies $a \wedge b \in \mathcal{F}$, for all a, b, and (iii) $a \in \mathcal{F}$ and $a \wedge b = a$ implies $b \in \mathcal{F}$, for all a, b. For example, $(\mathcal{P}(\mathbf{N}), \cup, \cap, \sim, \mathbf{N}, \emptyset)$ is a Boolean algebra, and the set of all cofinite subsets of $\mathbf{N}$ is a filter in this Boolean algebra.

(3) If $(P, \leq)$ is a partially ordered set, $\mathcal{F} \subseteq P$ is a *filter in* P if (i.) for all $a, b \in \mathcal{F}$, there exists $c \in \mathcal{F}$ such that $c \leq a$ and $c \leq b$, and (ii.) for all $a \in \mathcal{F}$ and all $x \in P$, $x \leq a$ implies $x \in \mathcal{F}$.

filter in a partial order   A nonempty subset G of a partial order $(P, \leq)$ such that any two elements of G are compatible in G and G is closed upwards. That is, for any p and q in G, there is an $r \in G$ with $r \leq p$ and $r \leq q$ (compatibility), and for any $p \in G$, if $q \in P$ with $p \leq q$, then $q \in G$ (closed up). Filters capture large elements in a partial order in the same way that ideals capture small elements.

final object   An object F in a category C with the property that, for any object X in C, there exists a unique morphism $g \in \mathrm{Hom}_C(X, F)$.

finite cardinal   A natural number, regarded as a cardinal number.

finite character   A set A (of sets) is *of finite character* if $A \neq \emptyset$ and for all sets X, X is in A if and only if every finite subset of X is in A.

For example, let A be the collection of all sets that contain pairwise disjoint subsets of $\mathbf{N}$. Then A has finite character.

finite continued fraction   A real number of the form

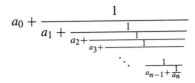

$$a_0 + \cfrac{1}{a_1 + \cfrac{1}{a_2 + \cfrac{1}{a_3 + \cfrac{\ddots}{\quad \cfrac{1}{a_{n-1} + \frac{1}{a_n}}}}}}$$

where each a_i is a real number. If each a_i is an integer, the fraction is said to be a *simple finite continued fraction*. It can be shown that a real number is rational if and only if it can be expressed as a finite (simple) continued fraction. For example,

$$\frac{10}{27} = \cfrac{1}{2 + \cfrac{1}{1 + \cfrac{1}{2 + \frac{1}{3}}}}$$

As an alternative to this cumbersome notation, the finite continued fraction decomposition of a number is often abbreviated $[a_0; a_1, a_2, \ldots, a_n]$, so that $\frac{10}{27}$ is denoted $[0; 2, 1, 2, 3]$.

finite intersection property   The property of a collection $\mathcal{C}$ of subsets of a set X that, for every finite subcollection $\{C_1, \ldots, C_n\}$ of $\mathcal{C}$, the intersection $C_1 \cap \cdots \cap C_n$ is non-empty.

finite ordinal   A natural number, regarded as an ordinal number. *See* ordinal.

finite set   A set that contains only finitely many elements. Equivalently, a set whose cardinality is a natural number.

first category   The class of topological space that is the countable union of nowhere dense subsets. Such a space is also called *meager*. *See* nowhere dense subset, second category space.

first countable space   A topological space X that has a countable basis at each point $x \in X$. That is, for each $x \in X$ there is a countable collection $\mathcal{B}_x$ of neighborhoods of x such that if $U \subset X$ is an open set containing x, then there is a set $B \in \mathcal{B}_x$ with $x \in B \subset U$.

first fundamental form   The quadratic form defined on tangent vectors to a surface M in Euclidean space $\mathbf{R}^3$ by taking the square of the length of the vector. If a portion of the surface

is parameterized by

$$X(u, v) = (x(u, v), y(u, v), z(u, v)),$$

then a tangent vector can be represented as a linear combination of the vectors $X_u = (\frac{dx}{du}, \frac{dy}{du}, \frac{dz}{du})$ and $X_v = (\frac{dx}{dv}, \frac{dy}{dv}, \frac{dz}{dv})$. Let

$$E(u, v) = X_u \cdot X_u, \quad F(u, v) = X_u \cdot X_v$$

and

$$G(u, v) = X_v \cdot X_v.$$

Then any tangent vector $a X_u + b X_v$ has length given by $\sqrt{Ea^2 + 2Fab + Gb^2}$. The first fundamental form is classically given in the form:

$$ds^2 = Edu^2 + 2Fdudv + gdv^2.$$

first order language   A *first order language* $\mathcal{L}$ for first order logic consists of the following alphabet of symbols:

(i.) $(,)$ (parentheses)

(ii.) $\neg, \rightarrow$ (logical connectives)

(iii.) an infinite list of variables $v_1, v_2, \ldots$

(iv.) a symbol $=$ for equality (which is optional)

(v.) a quantifier $\forall$

(vi.) predicate symbols: for each positive integer n, a particular, possibly empty, set of symbols, called n-place predicate symbols

(vii.) constant symbols: a particular, possibly empty, set of symbols, called constant symbols

(viii.) function symbols: for each positive integer n, a particular, possibly empty, set of symbols, called n-place function symbols (constant symbols are sometimes called 0-place function symbols).

In item (ii.), any complete set of logical connectives could be used. In item (v.), the universal quantifier $\forall$ could be replaced by the existential quantifier $\exists$.

Such a language is called a first order language because the quantifier ranges over variables only, as opposed to a second order language, where there are two types of quantifiers.

Some examples of first order languages are the language of set theory and the language of elementary number theory.

The language of set theory is a first order language that contains equality and one two-place predicate $\in$. In this language, the variables are intended to represent sets, and $\in$ is intended to be interpreted as "is an element of".

The language of elementary number theory is a first order language that contains equality, a single constant symbol 0, one two-place predicate $<$, one one-place function symbol S, and three two-place function symbols $+$, $\cdot$, and E. In this language, the variables are intended to represent natural numbers, S is intended to be interpreted as the successor function, and 0, $<$, $+$, $\cdot$, and E are intended to be interpreted as 0, the usual ordering on the natural numbers, addition, multiplication, and exponentiation, respectively.

first order logic   A formal logic with symbols from a first order language, rules that tell which expressions from the language are well-formed formulas, a semantic notion of truth (*see* structure, satisfy), and a syntactical notion of provability (*see* predicate calculus, proof).

First order logic is also called *predicate logic*.

fixed point   Let $f : X \to X$ be continuous. A point $x_0 \in X$ is a *fixed point* for f if $f(x_0) = x_0$. *See* Brouwer Fixed-Point Theorem.

focal property of a conic   A property of a conic section with regard to its focus or foci. For an ellipse, this property is lines drawn from the foci to a fixed point on the ellipse make equal angles with the tangent at the point. For a hyperbola, it is lines drawn from the foci to a fixed point on the hyperbola make an angle that is bisected by the tangent at the point. For a parabola, it is the line from the focus to a fixed point on the parabola makes an angle with the tangent equal to that made by the tangent with the line parallel to the axis of the parabola passing through the point.

focus   A point or points in the plane, corresponding to a given conic section, whose role varies depending upon the type of conic. An ellipse may be thought of as the locus of points in the plane whose distances from the foci have a constant sum. A hyperbola may be thought of as the locus of points in the plane whose distances from the foci have a constant difference. A parabola may be thought of as the locus of

points in the plane whose distances from the *focus* and a given line (*see also* directrix) are equal.

foliation   A family $\{N_\lambda : \lambda \in \Lambda\}$ of arcwise-connected pairwise disjoint subsets covering a given manifold M such that every point in M has a local coordinate system $(x^1, \ldots, x^n)$ so that each N_λ is given by $x^{n-k+1} =$ constant, $\ldots$, $x^n =$ constant for some $0 \le k \le n$.

foot of perpendicular   Suppose l is a line in the Euclidean (or hyperbolic) plane and P is a point not lying on l. Then the unique point Q lying on the line l such that the line through P and Q is perpendicular to l is called the *foot of the perpendicular* from P to l.

forgetful functor   A functor F from a category C to **Set** that assigns to each object $A \in \mathrm{Obj}(C)$ its underlying set (also denoted by A) and to each morphism $f : A \to B$ in C the function $f : A \to B$. Thus, the functor "forgets" any additional properties that the objects and morphisms in C have. For example, if $C = $ **Grp**, then a group $A \in C$ is mapped to the set A and a group homomorphism $f : A \to B$ is mapped to the function f; all group-theoretic properties possessed by A and f are ignored. *See* functor.

formal proof   *See* proof.

formal theory   *See* theory.

four-space   The topological vector space formed by taking the Cartesian product of four copies of the real line, denoted E^4, $\mathbf{R}^4$ or $\mathbb{R}^4$. A point in *four-space* is uniquely determined by an ordered quadruple (a, b, c, d) of real numbers.

fraction   If a and b are integers with $b \ne 0$, then the *fraction* $a \div b$ denotes the rational number resulting from the quotient $a \div b$.

Fréchet filter   *See* filter.

free variable   Let $\mathcal{L}$ be a first order language. If x is a variable and α is a well-formed formula of $\mathcal{L}$, then x occurs *free* in α (or x is a free variable in α) is defined by induction on complexity of α, as follows:

(i.) If α is an atomic formula, then x occurs free in α if x occurs in α.

(ii.) If $\alpha = (\neg\beta)$, then x occurs free in $(\neg\beta)$ if x occurs free in β.

(iii.) If $\alpha = (\beta \to \gamma)$, then x occurs free in $(\beta \to \gamma)$ if x occurs free in β or in γ.

(iv.) If $\alpha = \forall v_i \beta$, then x occurs free in $\forall v_i \beta$ if x occurs free in β and $x \neq v_i$.

As an example, v_1 and v_3 occur free, while v_2 does not occur free, in

$$\forall v_2(v_1 = v_2 \to \forall v_1(v_1 = v_3)).$$

Frenet frame   An orthonormal frame $\{T(t), N(t), B(s)\}$ of vectors at the point $C(s)$ on a given curve in $\mathbf{R}^3$, giving a moving coordinate system along the curve. Assume C has three continuous derivatives and that $C'(t)$ and $C''(t)$ are linearly independent. The first vector, T, is the unit vector tangent to the curve, given by $\frac{C'(t)}{||C'(t)||}$. The second unit vector, N, is the principal normal to the curve. It lies in the plane spanned by $C'(t)$ and $C''(t)$, is perpendicular to T, and is chosen so that it makes an acute angle with $C''(t)$. The third vector, B, is the binormal vector. It is defined by $B = T \times N$.

Frenet's formulas   Equations that relate the fundamental geometric invariants of a curve in Euclidean space or, more generally, in a Riemannian 3-manifold.

Suppose $C(s)$ is a curve possessing three continuous derivatives, parameterized by arc length. Assume that $C''(s) \neq 0$. Then the curve has a Frenet frame (T, N, B) satisfying the following linear system of differential equations:

$$C'(s) = T(s)$$

$$T'(s) = k(s)N(s)$$

$$N'(s) = -k(s)T(s) + \tau(s)B(s)$$

$$B'(s) = -\tau(s)N(s)$$

The function $k(s)$ is the geodesic curvature, and the function $\tau(s)$ is the torsion of the curve. *See* Frenet frame.

Frobenius integrability condition   The condition that must be satisfied by a k-dimensional distribution in an n-dimensional manifold in order for the distribution to be tangent to the leaves of a k-dimensional foliation. Given a manifold M, a distribution Δ assigns to each point P in M a k-dimensional subspace of the tangent space at P. It is integrable if the manifold is the union of k-dimensional submanifolds, such that the k-plane $\Delta(p)$ is the tangent plane of the k-manifold through p. The Frobenius condition says that if X and Y are vector fields defined in a neighborhood of P such that $X(Q)$ and $Y(Q)$ lie in $\Delta(Q)$, then the Lie bracket $[X, Y](Q)$ also lies in $\Delta(Q)$. *See* foliation.

Frobenius Theorem   A theorem that gives necessary and sufficient conditions for a distribution in a manifold to be tangent to the leaves of a foliation. Given a manifold M, a distribution Δ assigns to each point P in M a k-dimensional subspace of the tangent space at P. It is integrable if the manifold is the union of k-dimensional submanifolds, such that the k-plane $\Delta(p)$ is the tangent plane of the k-manifold through p. The *Frobenius Theorem* says that Δ is integrable if and only if, whenever X and Y are vector fields defined in a neighborhood of P such that $X(Q)$ and $Y(Q)$ lie in $\Delta(Q)$, then the Lie bracket $[X, Y](Q)$ also lies in $\Delta(Q)$.

frustrum   The portion of a cone lying between its base and a plane parallel to the base.

full subcategory   *See* subcategory.

function   If X and Y are sets, then a *function from X to Y* is a relation $f \subseteq X \times Y$ (often written $f : X \to Y$) with the property that $(x, y), (x, z) \in f$ implies $y = z$. It is standard to write $f(x) = y$ when $(x, y) \in f$. *See* relation.

function space   (1) Let X and Y be topological spaces. The *function space Y^X* is the set of all continuous maps from X into Y. This space can be given several topologies, the most common being the *compact-open topology*. *See* compact-open topology.

(2) Any topological space whose elements are functions on some common domain.

functor   Either a covariant functor or a contravariant functor. If no description is specified, then the *functor* is assumed to be covariant. *See* covariant functor.

fundamental cycle   If M is a compact, orientable manifold of dimension n, then the n-dimensional homology of M is an infinite cyclic group. A generator of $H_n(M)$ is a *fundamental cycle*. If M is a polyhedral manifold, then the fundamental cycle in simplicial homology can be given by the n-chain which is the sum of the n-simplices. *See* homology group.

Fundamental Theorem of Arithmetic   If n is an integer greater than 1, then n is either a prime number or can be expressed as a product of prime numbers, uniquely, except for order. For example, $24 = 2 \times 2 \times 2 \times 3$ and $30 = 2 \times 3 \times 5$.

Fundamental Theorem of the Theory of Curves   A curve in Euclidean space $\mathbf{R}^3$ is uniquely determined up to rigid motion by its geodesic curvature κ and torsion τ, as functions of its arc length parameter s. More precisely, given two continuous functions $\kappa(s)$ and $\tau(s)$ of one real variable such that $\kappa > 0$, and given initial values $X(0)$ and $X'(0)$ in $\mathbf{R}^3$ with $|X'(0)| = 1$, there is a unique curve $X(s)$ whose curvature is κ and torsion is τ. κ is usually taken to be continuously differentiable and X then has three continuous derivatives.

Fundamental Theorem of the Theory of Surfaces   Let S be a surface in Euclidean three-space parameterized by

$$X(u, v) = (x(u, v), y(u, v), z(u, v)) ,$$

where x, y, and z are assumed to have continuous third-order partial derivatives. Then S possesses a first fundamental form g and a second fundamental form L satisfying the Gauss equations and the Codazzi-Mainardi equations. The fundamental theorem of the theory of surfaces states the converse, namely, if $g(u, v)$ is a positive definite symmetric tensor (i.e., an inner product), and $L(u, v)$ is a symmetric tensor, with g having continuous second derivatives and L having continuous first derivatives, and if g and L satisfy the Gauss and Codazzi-Mainardi equations, then they (locally) determine a surface S uniquely up to rigid motion. *See* first fundamental form, Gauss equations, Codazzi-Mainardi equations.

G

G$_\delta$ set   A countable intersection of open sets. *See* F$_\sigma$ set.

Gauss equations   A system of partial differential equations arising in the theory of surfaces. If M is a surface in $\mathbf{R}^3$ with local coordinates (u^1, u^2), its geometric invariants can be described by its first fundamental form $g_{ij}(u^1, u^2)$, and second fundamental form $L_{ij}(u^1, u^2)$. The Christoffel symbols Γ^k_{ij} are determined by the first fundamental form. In order for functions g_{ij} and L_{ij}, $i, j = 1, 2$ to be the first and second fundamental forms of a surface, certain integrability conditions (arising from equality of mixed partial derivatives) must be satisfied. One set of conditions, the *Gauss equations,* relate the determinant of the second fundamental form to an expression involving only the first fundamental form (and its first and second partial derivatives). *See also* Christoffel symbols, first fundamental form.

general Cantor set   A Cantor set in which intervals of length $\frac{\alpha}{3^n}$ are removed at stage n for $0 < \alpha < 1$. The resulting set, $C_\alpha = \cap_{n=1}^{\infty} I_n$, is a closed set with length $1 - \alpha$ which forms a totally disconnected compact topological space in which every element is a limit point of the set. *See* Cantor set.

generalized continuum hypothesis   The statement

$$2^{\aleph_\alpha} = \aleph_{\alpha+1}$$

for all ordinals α. (Both this statement and its negation are independent of the axioms of Zermelo-Fraenkel set theory with the Axiom of Choice.) *See* continuum hypothesis.

generalized Riemann hypothesis   An assertion concerning the zeros of functions that are similar to the Riemann zeta function. These functions are called L-functions and are defined

as

$$L(s, \chi) = \sum_{n=1}^{\infty} \frac{\chi(n)}{n^s}$$

where χ is a *Dirichlet character*, that is, a real or complex valued function defined on the positive integers so that
(i.) $\chi(mn) = \chi(m)\chi(n)$ for all m and n;
(ii.) there is a positive integer k so that $\chi(n + k) = \chi(n)$ for all n;
(iii.) $\chi(n) = 0$ if $\gcd(n, k) > 1$.
The *generalized Riemann hypothesis* conjectures that all zeros of the function $L(s, \chi)$ on the "critical strip" (those complex numbers $s = x + iy$ such that $0 < x < 1$) must lie on the "critical line" ($x = \frac{1}{2}$). Notice that $L(s, 1) = \zeta(s)$, so the generalized Riemann hypothesis "agrees with" the Riemann hypothesis. *See also* Riemann zeta function, Riemann Hypothesis.

General Position Theorem   A name given to a number of theorems asserting, for various classes of maps of spaces X into spaces Y, that such maps may be approximated by maps having simpler structure. These "generic" maps usually have minimum possible dimensionality of intersections or self-intersections. A simple example is the statement that two affine subspaces of dimensions k and l in n-dimensional space can be perturbed so that they intersect in a subspace of dimension $n - (k + l)$.

generating curve   A surface of revolution in Euclidean space can be parameterized by

$$X(t, \theta) = (r(t)\cos(\theta), r(t)\sin(\theta), z(t)),$$

where $r(t) > 0$ and $z(t)$ are continuous functions. The curve $\sigma(t) = (r(t), z(t))$ is the *generating curve* of the surface.

geodesic   In a manifold with metric (nondegenerate smoothly varying quadratic form on each tangent space), a curve of minimal length between two points. An example of *geodesics* are arcs of great circles on a sphere.

geodesic correspondence   A smooth map $\phi : S \longrightarrow S'$ between two surfaces which takes geodesics of S to geodesics of S'. An isometric mapping is automatically a *geodesic correspondence,* but the converse is not true.

An example of a geodesic correspondence is given by central projection of the hemisphere $z = \sqrt{1 - x^2 - y^2}$ to the plane $z = 1$. Great circle arcs in the hemisphere correspond to straight lines in the plane under this projection.

geodesic curvature   If C is a curve lying on a surface S in Euclidean space $\mathbf{R}^3$, parameterized by unit speed, then its second derivative or acceleration vector C'' is perpendicular to the velocity vector C'. The length of the component of C'' tangent to the surface is the *geodesic curvature* of the curve. It measures the amount of bending the curve undergoes within the surface, as opposed to the amount of bending due to the bending of the surface itself.

geometric realization   Let S be an abstract simplicial complex. If S is isomorphic to the vertex scheme of a simplicial complex K, then K is the *geometric realization* of S. The geometric realization of an abstract simplicial complex is unique up to isomorphism.

geometry on a surface   The measurement of lengths of curves, angles between curves, and areas of figures lying on a surface. This is also called *intrinsic geometry*, to distinguish it from properties of a surface that depend on how the surface sits in space.

Gödel number   A *Gödel numbering* (arithmetization) is an effective method of coding nonnumerical objects by natural numbers. For example, there is a computable bijection from the set $\mathcal{T}$ of all Turing machine programs to the set of natural numbers. The inverse of this bijection is also computable, so that, given a Turing machine program, one can effectively find the natural number assigned to it, which is called the *Gödel number* of the program, and given a natural number, one can effectively "decode" it in order to find the Turing machine program that corresponds to it. The notation φ_e is used to denote the Turing computable partial function with Gödel number e; i.e., φ_e is Turing computable via the Turing program with Gödel number e. Gödel was the first to use Gödel numbers in his proof of his Incompleteness Theorem.

Gödel set theory   The same as Bernays-Gödel set theory. *See* Bernays-Gödel set theory.

great circle   A circle on a sphere formed by intersecting the sphere with a plane that passes through the center of the sphere. If P and Q are two points on the sphere, then a curve of least length joining P to Q is an arc of a *great circle*. *See also* geodesic.

greatest common divisor   (1) If a and b are nonzero integers, then the *greatest common divisor* of a and b, denoted $\gcd(a, b)$ is the largest integer that is a divisor of both a and b. For example, the greatest common divisor of 28 and 36 is 4 (the common divisors of 28 and 36 are ± 1, ± 2, and ± 4).

(2) Alternatively, in a Euclidean ring, R, $\gcd(a, b)$ is an element (not necessarily unique) d of R satisfying

(i.) d is a divisor of both a and b;

(ii.) if x is a divisor of a and b, then x is also a divisor of d.

greatest element   Given a set A with an ordering $\leq$ on A, an element $u \in A$ is said to be a *greatest element* of A if, for all $x \in A$, $x \leq u$. Note that if A has a greatest element, then it is unique. *Compare with* least element.

greatest lower bound   Let A be an ordered set and let $B \subseteq A$. An element $l \in A$ is said to be a *greatest lower bound* (or *infimum*) for B if it is a lower bound for B (i.e., for all $x \in B$, $l \leq x$) and if it is the greatest element in the set of all lower bounds for B (i.e., for all $y \in A$, if for all $x \in B$, $y \leq x$, then $y \leq l$). Note that if a set has a greatest lower bound, then it is unique. *Compare with* least upper bound.

group   A non-empty set G with a product map $G \times G \longrightarrow G$ ((g, h) is taken to gh), an inverse map $G \longrightarrow G$ (g is taken to g^{-1}), and a distinguished element called the identity (often denoted 0, 1, or e) satisfying $ge = eg = g$, for $g \in G$. These satisfy the relationships that $g_1(g_2 g_3) = (g_1 g_2)g_3$ and $gg^{-1} = e = g^{-1}g$.

Common examples include topological groups and Lie groups.

group of motions   A group of length-preserving transformations, or rigid motions, in Euclidean n-space. A group is a non-empty set with a binary associative operation that contains an identity and an inverse of each one of its elements. The rigid motions of the Euclidean plane are translations, rotations, reflections, and glide-reflections.

group of symmetries   Of a figure, a group of motions that transform a figure into itself. *See* group of motions. For example, the group of symmetries of an equilateral triangle is the group of six elements that can be identified with the permutations of the vertices.

H

half line   A connected unbounded and proper subset of a line in Euclidean space. Also called a ray. *See also* closed half line.

half plane   (1) (Open) One of the two connected sets remaining after deleting a line from a plane.

(2) (Closed) An open half plane, together with the deleted line. Example: On the Cartesian plane an open half plane is the set $\{(x, y) : ax + by + c > 0\}$ where $(a, b) \neq (0, 0)$, a and c are constant, the corresponding closed half plane is the set $\{(x, y) : ax + by + c \geq 0\}$.

half space   One of the two connected sets remaining after deleting from a sphere its intersection with a plane through the center.

halting problem   Informally, the *halting problem* asks if there is an effective procedure which, given an arbitrary effective procedure and input a natural number n, answers "yes" if that program on input n halts, and outputs "no" otherwise.

Formally, the halting set K_0 is the set of codes of pairs of natural numbers (e, x) such that the partial recursive function with Gödel number e is defined on input x; i.e., $K_0 = \{\langle e, x \rangle : \varphi_e(x)$ is defined$\}$, where φ_e is the partial recursive function with Gödel number e. The halting set is not recursive (computable), as a set of natural numbers, although it is recursively (computably) enumerable. Another halting set is $K = \{e : \varphi_e(e)$ is defined$\}$. The set K is also computably (recursively) enumerable but non–computable (non–recursive).

Hausdorff Maximal Principle   Every chain in a partially ordered set can be extended to a maximal chain. This principle is provable in ZFC, and it is provably equivalent to the Axiom of Choice and Zorn's Lemma in ZF.

Hausdorff metric   Let (X, d) be a metric space. If $A \subset X$ and $\epsilon > 0$, let $U(A, \epsilon)$ be the ϵ-neighborhood of A. That is,

$$U(A, \epsilon) = \cup_{a \in A} B(a, \epsilon)$$

where $B(a, \epsilon) = \{x \in X : d(x, a) < \epsilon\}$. Let $\mathcal{H}$ denote the collection of all (non-empty) closed, bounded subsets of X. If $A, B \in \mathcal{H}$ then the *Hausdorff Metric* on $\mathcal{H}$ is defined by $D(A, B)$

$$= \inf\{\epsilon : A \subset U(A, \epsilon) \text{ and } B \subset U(B, \epsilon)\}.$$

Hausdorff topological space   A topological space X such that, for each pair of distinct points $x, y \in X$, there exist open neighborhoods U and V of x and y, respectively, such that $U \cap V = \emptyset$. Also called T_2-space. See separation axioms.

height (of a tree)   The least ordinal α such that the αth level of a given tree T is empty. That is, α is the first ordinal for which there is no element in T whose predecessors have order type α. For example, if $T = \{\{a\}, \{a, b\}\}$ ordered by inclusion, then $\text{Lev}_0(T)$ contains the set $\{a\}$, $\text{Lev}_1(T)$ contains $\{a, b\}$, and the height of T is two.

Equivalently, the height of T may be defined by

$$\text{height}(T) = \sup_{t \in T}\{\text{ordertype}(\{s \in T : s < t\}) + 1\}.$$

heptagon   A plane polygon with seven sides. A (convex) *heptagon* is called regular when its sides have equal length. In that case, its vertices lie on a circle and all of the edges joining two neighboring ones are of equal length; for example, the vertices $\left(\cos\frac{2\pi i}{7}, \sin\frac{2\pi i}{7}\right)$, $i = 0, \ldots, 6$.

hereditary property of a topological space   A property $\mathcal{P}$ of a topological space X such that every subspace $A \subseteq X$ has property $\mathcal{P}$. For example, the property of being Hausdorff is hereditary, while the property of being compact is not.

hexagon   A plane polygon with six vertices.

hexahedron   A polyhedron with six faces. The most familiar regular *hexahedron* is the cube,

1-58488-050-3/01/$0.00+$.50
© 2001 by CRC Press LLC

regular for a (convex) polyhedron meaning that all the faces are equal regular polygons and all the vertices belong to the same number of faces.

Hilbert cube The Cartesian product $\prod_{n=1}^{\infty} I$ of countably many closed unit intervals. It is homeomorphic to $\prod_{n=1}^{\infty} \left[0, \frac{1}{n}\right]$ as well as the subspace

$$H = \{(x_n) \in \mathbf{R}^{\infty} : \sum_{n=1}^{\infty} x_n^2 < \infty\}.$$

holomorphic function A function

$$f(z_1, \ldots, z_n)$$

that is equal to the sum of an absolutely convergent power series in a suitable polydisc near each point of its domain (the radius may depend on the point):

$$f(z_1, \ldots, z_n) = \sum_{a_j \geq 0} c(a_1, \ldots, a_n) z_1^{a_1} \ldots z_n^{a_n},$$

$c(a_1, \ldots, a_n) \in \mathbf{C}$. When $n = 1$, this condition is equivalent to the Cauchy-Riemann equations:

$$\frac{\partial u}{\partial x} = \frac{\partial v}{\partial y}, \frac{\partial v}{\partial y} = -\frac{\partial v}{\partial x}$$

for $z = x + iy$, $f(z) = u(x, y) + iv(x, y)$. Examples include polynomials in z and the exponential function e^z.

holomorphic local coordinate system For a complex analytic manifold of dimension n, a biholomorphic identification $\phi_{U(p)}$ of a suitable open neighborhood $U(p)$ of each point p with the open ball of radius 1 and center the origin in $\mathbf{C}^n$, $\phi_{U(p)} : B_0 \to U(p)$.

homogeneous topological space A topological space X such that, for each pair of points x and y in X, there is a homeomorphism $h : X \to X$ such that $h(x) = y$.

homology class of a map Let $f : X \to Y$ be continuous and let $\simeq$ denote the homotopy equivalence relation. The homotopy class of f is the equivalence class

$$[f] = \{g : X \to Y : g \text{ is continuous and } g \simeq f\}.$$

If $A \subseteq X$ is a subspace, then the homotopy class of f rel A is the equivalence class $[f]_A$

$$= \{g : X \to Y : g \text{ is continuous and } g \simeq f \text{ rel } A\}.$$

homology equivalence Let $\mathcal{T}$ denote the category of all pairs (X, A) of topological spaces, where X is identified with the pair $(X, \emptyset)$. A *homotopy equivalence* between the pairs (X, A) and (Y, B) is a pair of functions $f : (X, A) \to (Y, B)$ and $g : (Y, B) \to (X, A)$, such that $g \circ f$ is homotopic to the identity map i_X of X and $f \circ g$ is homotopic to the identity map i_Y of Y.

homology group Let n be a positive integer, X a topological space, and x a point in X. Then the nth homotopy group $\pi_n(X, x)$ is defined to be the group of homotopy classes of maps of the standard sphere S^n to X taking a fixed base point $*$ of S^n to x. In the case $n = 1$, this is the fundamental group of X (with base point x), with concatenation of loops inducing the group operation. In dimensions higher than 1, the group operation turns out to be commutative. Another way of defining $\pi_n(X, x)$ is to take homotopy classes of maps of $[0, 1]^n$ to X taking the boundary to x. With this definition, it is easier to define the group operation by concatenation: $f + g : [0, 1]^n \longrightarrow X$ is defined by: $f + g(t_1, t_2, \ldots, t_n)$ equals $f(2t_1, t_2, \ldots, t_n)$ for $0 \leq t \leq \frac{1}{2}$, and it equals $g(2t_1 - 1, t_2, \ldots, t_n)$ for $\frac{1}{2} \leq t_1 \leq 1$.

homology theory A *homology theory* on a category $\mathcal{T}$ of pairs of topological spaces (X, A) consists of

(i.) a functor H_p from $\mathcal{T}$ to the category of Abelian groups $\mathcal{A}$ for each integer $p \geq 0$, where the image of the pair (X, A) is denoted by $H_p(X, A)$ and

(ii.) a natural transformation

$$\partial_p : H_p(X, A) \to H_{p-1}(A)$$

for each integer $p \geq 0$, where A denotes the pair $(A, \emptyset)$, which satisfy the Eilenberg-Steenrod Axioms. *See* Eilenberg-Steenrod Axioms.

homothety A transformation of the Euclidean plane to itself which takes every triangle to a similar triangle. The homotheties of the plane form a group under composition, which contains all isometries and the dilations given by $f(x, y) = (ax, ay), a \neq 0$.

homotopy type of a space Let $\mathcal{T}$ denote the category of all pairs (X, A) of topological spaces, where X is identified with the pair $(X, \emptyset)$. The *homotopy type of the pair* (X, A) is the equivalence class

$$[(X, A)] = \{(Y, B) : (Y, B)$$

is homotopy equivalent to $(X, A)\}$.

Hopf bundle The bundle $S^1 \longrightarrow S^3 \longrightarrow S^2$ formed as follows. Consider S^3 as the unit sphere in $\mathbf{C}^2$ (where $\mathbf{C}$ denotes the complex numbers). The sphere S^2 is given by $\mathbf{CP}^1$, the space of complex lines in $\mathbf{C}^2$, by identifying the line through the point (z_1, z_2) with the point z_2/z_1 in $\mathbf{C}$ when $z_1 \neq 0$ and identifying $\mathbf{C}$ with the sphere less one point; the complex line given by points $(0, z_2)$ is identified with the remaining point on S^2. The projection $S^3 \longrightarrow S^2$ is given by the map which sends (z_1, z_2) to the complex line through (z_1, z_2) (thought of as a point of $\mathbf{CP}^1$ identified with a point in S^2). Each fiber is homeomorphic to S^1.

Using quaternions and Cayley numbers instead of complex numbers, one can define analogous bundles $S^7 \longrightarrow S^4$ and $S^{15} \longrightarrow S^8$ with fibers S^3 and S^7, respectively. All three bundles are called *Hopf bundles*.

hyperbolic paraboloid One of the quadratic surfaces in $\mathbf{R}^3$. Since a symmetric matrix over the reals is congruent to one in diagonal form, the (non-degenerate) quadric surfaces are classified, by the sign of their eigenvalues and the configuration of their points at infinity into ellipsoids, hyperboloids, elliptic paraboloids, and *hyperbolic paraboloids,* the latter having canonical equation $\frac{2z}{c} = \frac{x^2}{a^2} - \frac{y^2}{b^2}$ for some non-zero constants a, b, c.

hyperbolic plane A plane satisfying the axioms of hyperbolic geometry, which comprise Hilbert's axioms of plane geometry and the "characteristic axiom of hyperbolic geometry": for

any line l and point p not on l, there are at least two lines on p not meeting l. A model of the *hyperbolic plane* is given by the unit disc $D = \{(x, y) \in \mathbf{R}^2 : x^2 + y^2 < 1\}$, with the Poincaré metric $ds^2 = \frac{dx^2 + dy^2}{(1 - (x^2 + y^2)/4)^2}$, the lines being geodesic.

hyperboloid One of the quadric surfaces in $\mathbf{R}^3$, like the hyperbolic paraboloid. *See* hyperbolic paraboloid. The canonical equation for the hyperboloid of one sheet is $\frac{x^2}{a^2} - \frac{y^2}{b^2} + \frac{z^2}{c^2} = 1$ and that for the *hyperboloid* of two sheets is $\frac{x^2}{a^2} - \frac{y^2}{b^2} - \frac{z^2}{c^2} = 1$, where a, b, c are non-zero constants. As the name suggests, one difference between the two cases is that one surface consists of one connected component, the other of two.

hyperelliptic surface A Riemann surface X, namely a compact complex manifold of dimension 1, which generalizes the complex torus, in the following sense: there exists a holomorphic map $X \rightarrow \mathbf{P}^1$ of degree 2, with $2g + 2$ branch points where $2 - 2g$ is the Euler characteristic of the topological surface X. *See* compact complex manifold, complex torus. The number g is called the genus of the surface. Also called elliptic curve, in the case of genus 1.

hyperplane In n-dimensional (affine or projective) space of dimension n, a subspace of dimension $n - 1$.

hyperplane at infinity In n-dimensional projective space $\mathbf{P}^n$, with given coordinate system $(x_0 : x_1 : \ldots : x_n)$, a hyperplane H, typically given by the equation $x_0 = 0$. The complement $\mathbf{P}^n \backslash H$ can thus be identified with affine n-space A^n with coordinates $(\frac{x_1}{x_0}, \ldots \frac{x_n}{x_0})$.

hypersurface In affine or projective space, a subset defined by one (nonzero) algebraic equation in the coordinates. A hyperplane is an example of *hypersurface,* which is defined by one linear equation. *See also* hyperplane.

hypotenuse The side of a right triangle opposite the right angle. It is the longest side of the triangle.

I

icosahedron A *polyhedron* with 20 faces. The *icosahedron* is one of the five (convex) polyhedra that can be *regular*.

ideal Let S be a nonempty set and let $\mathcal{P}(S)$ be the power set of S. A set $I \subseteq \mathcal{P}(S)$ is an *ideal* on S if
   (i.) $\emptyset \in I$
   (ii.) for all $X, Y \in I$, $X \cup Y \in I$,
   (iii.) for all X, Y, if $X \in I$ and $Y \subseteq X$, then $Y \in I$.

As an example, let S be the set $\mathbf{N}$ of natural numbers and let I be the set of all finite subsets of $\mathbf{N}$. Then I is an ideal on S.

identification map A continuous onto mapping $f : X \to Y$ such that the topology on Y is the identification topology; that is, U is open in Y if and only if $f^{-1}(U)$ is open in X. *See also* quotient map.

identification space An *identification space* of a topological space X is a set Y endowed with the topology induced by an onto mapping $f : X \to Y$. This topology (the identification topology) is given by: $U \subseteq Y$ is open if and only if $f^{-1}(U)$ is open in X. *See also* quotient space.

identification topology *See* identification space.

identity function The arithmetic function, denoted I, which has the value 1 when $n = 1$ and has the value 0 when $n > 1$, i.e., $I(n) = \lfloor \frac{1}{n} \rfloor$, the floor function applied to $\frac{1}{n}$. It is completely (and strongly) multiplicative. This function is the identity under Dirichlet multiplication. *See* arithmetic function, Dirichlet multiplication.

image Let $f : A \to B$ be a function, and let $x \in A$. The *image* of x under f is $f(x)$, the unique element of B to which x is mapped by f. Given a subset $C \subseteq A$, the image of C under f is

$$f(C) = \{f(x) : x \in C\}$$
$$= \{y \in B : (\exists x \in C)[f(x) = y]\}.$$

imaginary axis The y-axis, which corresponds to the purely imaginary numbers in the Argand diagram for the complex numbers, namely the identification of $x + iy \in \mathbf{C}$ with the point $(x, y) \in \mathbf{R}^2$.

imbedding A one-to-one continuous map $f : X \to Y$, between topological spaces, for which the restriction $f^* : X \to f(X)$ to its range is a homeomorphism. Here

$$f(X) = \{y \in Y : \exists x \in X\ (f(x) = y)\}$$

and f^* is required not only to be one-to-one, onto, and continuous, but to have a continuous inverse.

immersed submanifold The image $f(M)$ of an immersion $f : M \to N$ between two manifolds. Each point in M has a neighborhood on which f is an embedding. However, the map f need not be an embedding, so $f(M)$ need not be a manifold with the induced topology as a subset of N, even if f is globally 1-1. A simple example is given by the immersion of the open interval into the plane whose image is the figure 6. A more complicated example is given by viewing the torus as the quotient of the plane by the integer lattice. Then a line of irrational slope is mapped onto a dense subset of the torus, which is an *immersed submanifold*.

inaccessible cardinal (strongly) A cardinal κ which is uncountable, regular, and satisfies the condition $2^\alpha < \kappa$ for all $\alpha < \kappa$, i.e., κ is a strong limit cardinal. (Any strongly inaccessible cardinal is weakly inaccessible since a strong limit cardinal is a limit cardinal. The existence of strongly inaccessible cardinals cannot be proved in Zermelo-Fraenkel set theory with the Axiom of Choice.) *See* regular cardinal.

inaccessible cardinal (weakly) A cardinal which is uncountable, a regular cardinal, and a limit cardinal. (The existence of weakly inaccessible cardinals cannot be proved in Zermelo-Fraenkel set theory with the Axiom of Choice.

As seen in the definition of inaccessible cardinal (strongly), every strongly inaccessible cardinal is weakly inaccessible. [*See* inaccessible cardinal (strongly).] If one assumes the generalized continuum hypothesis, then the converse is true.) *See* regular cardinal, limit cardinal.

incenter of triangle   The center of the unique circle which can be inscribed in a given triangle. It is located at the intersection of the internal bisectors of the three vertices of the triangle.

incommensurable   Two line segments XY and $X'Y'$ such that there is no line segment AB with the property that each of XY and $X'Y'$ has length that is an exact (integer) multiple of the length of AB. That is, there is no unit of measure with respect to which both segments have integer length.

For example, the hypotenuse of an isosceles right triangle and a leg of the triangle are *incommensurable* because $\sqrt{2}$ is an irrational number.

incomparable (elements of a partial ordering)   If $(P, \leq)$ is a partially ordered set, $x, y \in P$ are *incomparable* if neither $x \leq y$, nor $y \leq x$.

incompatible (elements of a partial ordering)   *See* compatible (elements of a partial ordering).

inconsistent   Let $\mathcal{L}$ be a first order language and let Γ be a set of well-formed formulas of $\mathcal{L}$. The set Γ is *inconsistent* if there exists a well-formed formula α such that both α and $(\neg\alpha)$ are provable from Γ (i.e., both α and $(\neg\alpha)$ are theorems of Γ). If Γ is *inconsistent,* then in fact every formula is a theorem of Γ.

inconsistent axioms   A set of axioms such that there is a statement A such that both A and its negation are provable from the axioms. *See also* inconsistent. For example, in set theory, the axioms AC (the Axiom of Choice) and AD (the Axiom of Determinacy) are inconsistent, as the Axiom of Determinacy contradicts the Axiom of Choice.

indiscernible   A subset I of a model A is a *set of indiscernibles* if no first-order formula can distinguish between increasing sequences from I. More precisely, if $<$ is any linear order on I

and $n \in \mathbf{N}$, then for all $a_1 < a_2 < \cdots < a_n$ and $b_1 < b_2 < \cdots < b_n$ in I, $A \models \phi(\bar{a})$ if and only if $A \models \phi(\bar{b})$ for all L-formulas ϕ.

induction   One of two techniques used to prove that a given proposition P is true for all natural numbers. Let $P(n)$ denote the statement "P is true for the natural number n". The principle of weak *induction* states that if

(i.) $P(0)$

(ii.) $P(m)$ implies $P(m + 1)$ for any natural number M,

then $P(n)$ for all natural numbers n. The principle of strong induction states that if (i.) and (ii.) $P(0), P(1), \ldots, P(m)$ implies $P(m + 1)$ for any natural number m, then $P(n)$ is true for any natural number n. The proof technique in the principle of strong induction may be generalized to any well-ordered set W, giving the principle of transfinite induction.

inductive set   A set A such that $\emptyset \in A$, and for all sets x, if $x \in A$ then $x^+ \in A$, where $x^+ = x \cup \{x\}$ is the successor of the set x.

infimum   Let $(X, \leq)$ be a partially ordered set and suppose that $Y \subseteq X$. An element $z \in X$ is an *infimum*, or *greatest lower bound*, of Y (denoted inf(Y) or glb(Y)) if z is a lower bound for Y and $r \leq z$ for any other element r which is a lower bound for Y. *See* lower bound.

infinite continued fraction   A continued fraction that is not finite. *See* continued fraction, finite continued fraction.

infinite dimensional projective space   A space in projective geometry, which generalizes the n-dimensional projective space $\mathbf{P}^n$. Over a field k, the points of $\mathbf{P}^n$ can be coordinatized by $(n + 1)$-tuples $(x_0 : x_1 : \ldots : x_n)$ where $x_i \in k$ and at least one $x_i \neq 0$, up to the equivalence relation $(x_0 : \ldots : x_n) \sim (\lambda x_0 : \ldots : \lambda x_n)$ for $0 \neq \lambda \in k$. The classical examples are the real projective space $(k = \mathbf{R})$ and complex projective space $(k = \mathbf{C})$. The *infinite dimensional projective space* can be constructed as a direct limit

$$\lim_{\rightarrow} \mathbf{P}^n = \mathbf{P}^\infty ,$$

namely a collection of injections $\pi_i : \mathbf{P}^i \rightarrow \mathbf{P}^\infty$ with the property that $\pi_j = \pi_i \circ \rho_{ij}$, where

$\rho_{ij} : \mathbf{P}^j \to \mathbf{P}^i$ is the natural inclusion $(x_0 : \ldots : x_j) \mapsto (x_0 : \ldots : x_j : 0 : \ldots : 0)$ for $j < i$.

infinite Grassmann manifold   A Grassmann manifold is the set of all subspaces of a vector space V that have a given dimension k. When V is a real (or complex) vector space, this set is indeed a real (or complex) manifold. If the dimension of V is n, the dimension of the Grassmann manifold is $k(n - k)$. An *infinite Grassmann manifold* is a generalization of this object for V of infinite dimension, but for it to be a manifold, care must be taken that there exist local coordinates. Typically, the condition for a subspace W to be a point of an infinite Grassmann manifold is that it be commensurable to a fixed subspace H of V, in a suitable sense, either involving the dimensions of $(H + W)/H$ and $W/(H \cap W)$ or some more analytic properties.

infinite set   Any set that is not finite. Equivalently, an *infinite set* is a set whose cardinality is not a natural number. *See* finite set.

infinite Stiefel manifold   The *infinite Stiefel manifold* V_k of k-frames is the direct limit (union) of the spaces $V_{n,k}$ of k-frames in real or complex n-dimensional space. More precisely, let $F = \mathbf{R}^\infty$ (resp., $\mathbf{C}^\infty$) denote the vector space of infinite sequences $x = (x_1, x_2, \ldots)$ of real (resp., complex) numbers that have only finitely many nonzero terms. Then the Stiefel manifold $V_k(F)$ of k-frames in F is the open subset of F^k consisting of k-tuples of linearly independent vectors in F.

infinity   The symbol ∞ was first used by the English mathematician John Wallis (1616-1703) to denote *infinity*. While not a number itself, ∞ is usually used to denote a quantity that is larger than every number. There are a number of ways mathematicians have attempted to "quantify" infinity. For instance, a set S is finite if there is a nonnegative integer n and a bijection $f : S \to \{1, 2, 3, \ldots, n\}$ (that is, if S has n elements) and S is infinite, otherwise. Alternatively, a set is infinite if there is a bijection from the set to a proper subset of itself. In his theory of transfinite numbers, Georg Cantor dealt with various "sizes" of infinity, distinguishing between countably and uncountably infinite sets, for example.

initial object   An object I in a category $\mathcal{C}$ with the property that, for any object X in $\mathcal{C}$ there exists a unique morphism $f \in \operatorname{Hom}_{\mathcal{C}}(I, X)$.

initial ordinal   If α is an ordinal, let $|\alpha|$ denote the cardinality of $\{\tau : \tau < \alpha\}$. The *initial ordinal* corresponding to a fixed cardinal κ is the minimum ordinal α such that $|\alpha| = \kappa$. For example, ω is the initial ordinal corresponding to $\aleph_0$ even though there are infinitely many different ordinals whose cardinality is $\aleph_0$. *See* ordinal.

initial segment   A subset of a well-ordered set W which has the form $\{x \in W : x < w\}$, where $w \in W$.

injection   (1) A one-to-one function between two sets X and Y. *See* one-to-one function.

(2) The function $i : X \to Y$ between two sets X and Y, with $X \subseteq Y$, defined by $i(x) = x$.

See also function.

integer   An element of the set $\{\ldots, -4, -3, -2, -1, 0, 1, 2, 3, 4, \ldots\}$ consisting of all whole numbers. The set of all *integers* is usually denoted $\mathbf{Z}$ or $\mathbb{Z}$.

integral curvature   If S is a surface in Euclidean space and A is a measurable subset of S, then the *integral curvature* of A is the area of its image under the Gauss map to the unit sphere. That is, it is the area of the set of unit vectors that are outer normals to support planes of the surface at points of A. For smooth surfaces, it can be computed intrinsically by integrating the Gaussian curvature over the region A. For a polyhedron, the integral curvature is concentrated at the vertices.

interior angle   If P is a simple closed polygon enclosing a region R, then the *interior angle* at a vertex V is measured by the magnitude of the rotation that carries one edge of P adjacent to V to the other edge, the rotation performed within R.

interior of closed curve   The bounded component of the complement of a simple closed curve. By the Jordan Curve Theorem, the complement of the curve consists of exactly two connected components.

interior of polygon   The bounded component of the complement of a polygon P, which, as a curve, is closed and has no self-intersections. Any two points in the interior can be joined by a continuous curve that does not intersect P, while any two points in the other component of the complement of P (the exterior of the polygon) can be joined by a continuous curve that does not intersect P. But no point in the interior can be joined to any point in the exterior by a continuous curve that does not intersect P. This fact is the content of the Jordan Curve Theorem for polygons.

interior of polyhedron   A closed connected polyhedral surface in Euclidean space $\mathbf{R}^3$ has a complement consisting of two path-connected components. One of these two, the bounded component, is called the *interior of the polyhedron*.

intersection of sets   If X and Y are sets, then the *intersection* of X and Y, denoted $X \cap Y$, is the set consisting of all elements that are common to both X and Y. Symbolically, $X \cap Y = \{z : z \in X \text{ and } z \in Y\}$. More generally, if $\{X_\alpha\}_{\alpha \in \Gamma}$ is a family of sets, then the intersection $\bigcap_{\alpha \in \Gamma} X_\alpha$ is the set consisting of all elements that are common to all X_α. *See also* Boolean algebra, lattice.

inverse correspondence   *See* inverse function.

inverse function   Let X, Y be sets and suppose that $f \subseteq X \times Y$ is a function. If f is one-to-one ($f(x_1) = f(x_2)$ implies $x_1 = x_2$), then the *inverse function* f^{-1} is the unique function obtained by interchanging the coordinates in the ordered pairs belonging to f. Symbolically, $f^{-1} = \{(y, x) : (x, y) \in f\}$. It can be verified that $f^{-1}f(x) = x$ and $ff^{-1}(y) = y$ for all $x \in X$ and all y in the range of f. The function f^{-1} is also one-to-one and its inverse satisfies $(f^{-1})^{-1} = f$. If f is not one-to-one, then f^{-1} is not a function. *See* function.

inverse morphism   Suppose that C is a category. If $f \in \mathrm{Hom}_C(A, B)$ is an invertible function, then the *inverse morphism* f^{-1} is the unique morphism in $\mathrm{Hom}_C(B, A)$ satisfying

(i.) $ff^{-1} = 1_B$,

(ii.) $f^{-1}f = 1_A$.

inverse relation   Let X, Y be sets and suppose that $R \subseteq X \times Y$ is a relation. The *inverse relation* R^{-1} is the relation obtained by interchanging the coordinates of all ordered pairs in R. Symbolically, $R^{-1} = \{(y, x) \in Y \times X : (x, y) \in R\}$. *See* relation.

involute   A curve associated with a given curve C as follows: the tangent lines to the curve form a surface, an *involute* is a curve on this surface which is orthogonal to the tangent lines. If C is parameterized by arc length s, then the involutes are given by $I_c(s) = C(s) + (c - s)C'(s)$, c a constant.

involution   A transformation that is its own inverse. In geometry, the reflection across a straight line is an *involution* of the plane.

irrational number   A real number that is not rational. That is, a real number that cannot be expressed as a quotient of integers. Examples of *irrational numbers* are π, e (the base of the natural logarithm), $\sqrt{2}$, and $\sqrt{6}$ (in fact, the square root of any integer, other than a perfect square, will be irrational).

irreducible quadratic polynomial   The polynomial $ax^2 + bx + c$ with real coefficients is *irreducible* (over the field of real numbers) if it cannot be expressed as the product of two non-constant polynomials with real coefficients. This occurs if and only if $b^2 - 4ac < 0$.

Any polynomial with real coefficients can be factored (using only real coefficients) into the product of linear factors and irreducible quadratic polynomials.

irreflexive relation   A relation $R \subseteq X \times X$ on a set X such that there is no $x \in X$ with $(x, x) \in R$. For example, if R consists of all ordered pairs of real numbers (a, b) such that $a < b$, then R is irreflexive.

isolated point   A point x in a topological space X such that the singleton set $\{x\}$ is open in X. Equivalently, $x \notin \overline{X \setminus \{x\}}$. Thus, x is isolated in X if and only if it is not an accumulation point in X.

More generally, x is an *isolated point* of a subset $A \subseteq X$ if $x \in A$ and there is an open $U \subseteq X$ with $U \cap A = \{x\}$. That is, $x \notin \overline{A \setminus \{x\}}$, and so x is an isolated point of A if and only if it is in A but is not an accumulation point of A.

isometric surfaces   Two surfaces S and S', for which there is a bijection from S to S', which takes every curve in S to a curve in S' of the same length. Assuming the surfaces are differentiable, they are isometric if there is a diffeomorphism from S to S' which pulls the first fundamental form of S' back to the first fundamental form of S. *See* first fundamental form.

isomorphic orderings   Two orderings $(X, \leq)$ and $(X', \leq')$ such that there exists a bijection from X to X' which is order-preserving. More precisely, the orderings are isomorphic if there is a bijection $f : X \to X'$ such that if $x_1, x_2 \in X$ and $x_1 \leq x_2$, then $f(x_1) \leq' f(x_2)$.

isomorphism   Let $\mathcal{L}$ be a first order language, and let $\mathcal{A}$ and $\mathcal{B}$ be structures for $\mathcal{L}$, where A and B are the universes of $\mathcal{A}$ and $\mathcal{B}$, respectively. A function $h : A \to B$ is an *isomorphism* of structures if h is injective and surjective, and

(i.) for each n-ary predicate symbol P and every $a_1, \ldots a_n \in A$,

$$(a_1, \ldots, a_n) \in P^{\mathcal{A}}$$
$$\Leftrightarrow (h(a_1), \ldots, h(a_n)) \in P^{\mathcal{B}},$$

(ii.) for each constant symbol c,

$$h(c^{\mathcal{A}}) = c^{\mathcal{B}},$$

and

(iii.) for each n-ary function symbol f and every $a_1 \ldots, a_n \in A$,

$$h(f^{\mathcal{A}}(a_1, \ldots, a_n)) = f^{\mathcal{B}}(h(a_1), \ldots, h(a_n)).$$

If there is an isomorphism of $\mathcal{A}$ onto $\mathcal{B}$, then $\mathcal{A}$ and $\mathcal{B}$ are *isomorphic structures* (notation: $\mathcal{A} \cong \mathcal{B}$).

isoperimetric   For two curves, C and C', the property of having the same length. The *isoperimetric* inequality in the plane states that, among all curves isoperimetric to a given simple closed curve C, the circle encloses the maximum area.

isosceles   An *isosceles polygon* is a polygon possessing two sides of the same length. The term is usually applied to triangles or trapezoids. In the case of a trapezoid, the sides are generally taken to be opposite sides.

isosceles triangle   A triangle possessing two sides of equal length. *See also* isosceles.

J

jump   Let A be a set of natural numbers. The *jump* of A (also called the *Turing jump* of A) is the halting set, relativized to A; i.e., the jump of A is the set $\{e : \varphi_e^A(e) \text{ is defined}\}$, where φ_e^A denotes the partial A–computable (A–recursive) function with Gödel number e. The jump of A is denoted by A'.

K

K3 surface   An algebraic surface that is smooth, has a global holomorphic 2-form, and first homology group of rank 0. Part of an important class of surfaces in algebraic geometry, named after three mathematicians: Kummer, Kähler, and Kodaira. An example is the intersection of three generic quadric hypersurfaces in $\mathbf{P}^5$. *See* hypersurface.

Kirby calculus   A method of specifying surgery operations on a manifold in terms of the identifications to be performed on the meridian of a solid torus embedded in the manifold.

Kleene's hierarchy   Alternate (rarely used) terminology for the arithmetical hierarchy. *See* arithmetical hierarchy.

Kleinian group   A subgroup G of the group of Möbius transformations, with the property that there exists some point z of the extended plane $\mathbf{C} \cup \{\infty\}$ at which G acts discontinuously, i.e., the stabilizer G_z is finite, and there exists a neighborhood U of z which is fixed by all the elements of G_z, but whose only fixed point under any element of G_z is z. *See* linear fractional function. Examples are given by the first homotopy group of Riemann surfaces.

Kodaira dimension   A rational invariant of a smooth projective variety V, named after the Japanese mathematician Kunihiko Kodaira. It is the maximum of the dimensions of $\phi_n(V)$, where ϕ_n is the rational map associated to the nth power of the canonical bundle, over all positive integers n for which this power has global sections. If no such n exists, the *Kodaira dimension* is defined to be $-\infty$.

k-perfect number   A positive integer n having the property that the sum of its positive divisors is kn, i.e., $\sigma(n) = kn$. Thus, a 2-perfect number is the same as a perfect number. The smallest 3-perfect number is 120. The smallest 4-perfect number is 30,240.

Kurepa tree   A tree of height ω_1 with no uncountable levels but at least ω_2 uncountable branches. Thus, for each $\alpha < \omega_1$, the α-level of T, $\mathrm{Lev}_\alpha(T)$, given by

$$\{t \in T : \mathrm{ordertype}(\{s \in T : s < t\}) = \alpha\}$$

is countable, $\mathrm{Lev}_{\omega_1}(T)$ is the first empty level of T, and there are at least ω_2 different sets $B \subseteq T$ totally ordered by $<$ (branches) that are uncountable. Kurepa's Hypothesis (KH) is that there exists a *Kurepa tree,* but KH is independent of the axioms of set theory. In fact, KH is independent of ZFC+GCH.

For any regular cardinal κ, a κ-Kurepa tree is a tree of height κ in which all levels have size less than κ and there are at least κ^+ branches with length κ. *See also* Aronszajn tree, Suslin tree.

L

Latin square   An $n \times n$ array of numbers such that each row and column of the array contains the same numbers and each number appears exactly once in every row and column. For example,

$$
\begin{array}{ccc}
1 & 2 & 3 \\
2 & 3 & 1 \\
3 & 1 & 2
\end{array}
\quad \text{and} \quad
\begin{array}{ccc}
1 & 2 & 3 \\
3 & 1 & 2 \\
2 & 3 & 1
\end{array}.
$$

lattice   A non-empty set X, together with two binary operations $\cup, \cap$ on X (called union and intersection, respectively), which satisfy the following conditions for all $A, B, C \in X$:

(i.) $(A \cup B) \cup C = A \cup (B \cup C)$;
(ii.) $(A \cap B) \cap C = A \cap (B \cap C)$;
(iii.) $A \cup B = B \cup A$;
(iv.) $A \cap B = B \cap A$;
(v.) $(A \cup B) \cap A = A$;
(vi.) $(A \cap B) \cup A = A$.

leaf   A manifold that is a maximal integral submanifold of an integrable distribution. Given a manifold M, a distribution Δ assigns to each point P in M a k-dimensional subspace of the tangent space at P. It is integrable if the manifold is the union of k-dimensional immersed submanifolds, such that the k-plane $\Delta(p)$ is the tangent plane of the k-manifold through p. A *leaf* is a maximal connected integral submanifold of the distribution.

least common multiple   For two nonzero integers a and b, the smallest positive integer L that is a multiple of both a and b, is denoted $\mathrm{LCM}(a, b)$. Equivalently, $\mathrm{LCM}(a, b)$ is the unique positive integer that is a multiple of both a and b and is a divisor of all other common multiples of a and b. For example, $\mathrm{LCM}(14, 8) = 56$ and $\mathrm{LCM}(3, 5) = 15$. Note that the *least common multiple* of two nonzero integers will always be a divisor of the product of the two integers. In fact, the product of the greatest common divisor and the least common multiple of two integers is the product of the two integers (i.e., $ab = \gcd(a, b) \cdot \mathrm{LCM}(a, b)$.) *See* greatest common divisor.

least element   Given a set A with an ordering $\leq$ on A, an element $l \in A$ is said to be a *least element* of A if, for all $x \in A$, $l \leq x$. Note that if A has a least element, then it is unique. *Compare with* greatest element.

least upper bound   Let A be an ordered set and let $B \subseteq A$. An element $u \in A$ is said to be a *least upper bound* (or *supremum*) for B if it is an upper bound for B (i.e., for all $x \in B$, $x \leq u$) and it is the least element in the set of all upper bounds for B (i.e., for all $y \in A$, if for all $x \in B$, $x \leq y$, then $u \leq y$). Note that if a set has a least upper bound, then it is unique. *Compare with* greatest lower bound.

left adjoint functor   Let C and D be categories with functors $F : C \longrightarrow D$ and $G : D \longrightarrow C$ such that if X is an object of C and Y is an object of D, we have a bijection of hom-sets

$$
\hom_C(X, G(Y)) = \hom_D(F(X), Y)
$$

which is natural in both X and Y. Then F is a left adjoint for G and G is a right adjoint for F.

Example: The forgetful functor from Abelian groups to sets which forgets the group structure has left adjoint given by taking the free Abelian group on the elements of the set. Note that this is *not* an "inverse" functor.

leg   In a right triangle, either of the two sides incident to the right angle.

level (of a tree)   The αth *level* of a tree T, $\mathrm{Lev}_\alpha(T)$, is the set of all elements of T whose predecessors have order type α. That is, for any $t \in T$, the set of predecessors of t, $\{s \in T : s < t\}$, must be well ordered, and the level of t is given by their order type.

$\mathrm{Lev}_0(T)$ is the set of elements in T with no predecessors, while $\mathrm{Lev}_1(T)$ is the set of elements with exactly one predecessor (which must come from level 0).

Lie group   A group that is also a differentiable manifold and for which the product and inverse maps are infinitely differentiable (and is therefore a topological group). *See* topological group.

Example: Consider the unit circle in the complex numbers, all points of the form e^{ix} for x real. This is a *Lie group.*

Lie line-sphere transformation   A correspondence between lines in space R, say, and spheres in a corresponding space S, named after the Norwegian mathematician Marius Sophus Lie. A point $(X, Y, Z) \in S$ determines a line in R by the two equations:

$$\begin{cases} (X + iY) - zZ - x = 0 \\ z(X - iY) + Z - y = 0 \end{cases}$$

For any fixed line l in R, the set of such lines that meet l corresponds to a sphere in S (whose center and/or radius may be complex numbers).

limit cardinal   A cardinal $\aleph_\alpha$ whose index α is a limit ordinal. *See* limit ordinal.

limit ordinal   An ordinal α that is not a successor ordinal. Therefore, α has the form $\sup\{\beta : \beta < \alpha\} = \bigcup_{\beta < \alpha} \beta$. (It should be noted that 0 is also a *limit ordinal;* we define $\sup \emptyset = 0$.) *See* successor ordinal, ordinal.

line bundle   A term used in the theory of vector bundles. A vector bundle over a topological space X consists of a space E called the total space, a vector space F called the fiber, and a map $\pi : E \longrightarrow X$. The space X has a covering by open sets U_i with homeomorphisms ϕ_i from $E_i = \pi^{-1}(U_i)$ to $U_i \times F$. The projection map π respects these product structures; i.e., $\pi \circ \phi_i^{-1}(x, V) = x$. When $x \in U_i \cap U_j$, the map $g_{ij}(x) : F \longrightarrow F$ defined by $(x, g_{ij}(x)(V)) = \phi_j(\phi_i^{-1}(x, V))$ is required to be linear. This implies that $\pi^{-1}(x)$ has the structure of a vector space. A *line bundle* is a vector bundle with one-dimensional fiber. Usually the field is either the real numbers or the complex numbers.

line of curvature   A curve C on a surface, having the property that, at each point $C(t)$ on the curve, the tangent vector $C'(t)$ is a principal vector of the surface at $C(t)$, that is, an eigenvector of the Weingarten map.

line segment   All points P on the line determined by two given points A and B, lying between A and B in the plane. For such a point P, the points A and B lie on different rays from P. This definition also makes sense in hyperbolic plane geometry but not in elliptic geometry, where betweenness is not a well-defined concept.

linear fractional function   A bijection of the extended complex plane $\mathbf{C} \cup \{\infty\}$ defined by $z \mapsto \frac{az+b}{cz+d}$ for given complex numbers a, b, c, d such that $ad - bc \neq 0$. The inverse is given by $z \mapsto \frac{dz-b}{cz-a}$. Also called Möbius transformation, linear fractional transformation, linear transformation.

linearly ordered set   A set A with a linear ordering on A. *See* linear ordering.

linear ordering   A partial ordering $\leq$ on a set A in which every pair of distinct elements of A is comparable; i.e., for all $x, y \in A$, if $x \neq y$, then $x \leq y$ or $y \leq x$. If the partial ordering is of the $<$ type (*see* partial ordering), then $<$ is a *linear ordering* if, for all $x, y \in A$, if $x \neq y$, then $x < y$ or $y < x$.

The usual ordering $\leq$ on $\mathbf{Q}$, the set of rational numbers, is a linear ordering.

A linear ordering is also called a total ordering.

link   A *link of n components* in $\mathbf{R}^3$ is a subset of $\mathbf{R}^3$ which is homeomorphic to n distinct copies of S^1. Individual components may be knotted with themselves or with other components.

More generally, a link of n components in $\mathbf{R}^{m+2}$ is an m embedding of a finite number of copies of S^m in $\mathbf{R}^{m+2}$ or S^{m+2}. Two links L_1, L_2 are equivalent if there is a homeomorphism $h : \mathbf{R}^{m+2} \to \mathbf{R}^{m+2}$ (or $h : S^{m+2} \to S^{m+2}$) such that $h(L_1) = L_2$.

linking number   A numerical invariant of links in 3 space which measures the number of times pairs of components of a link wrap about each other. *See* link.

For a link of 2 components $L = \alpha \cup \beta$ in $\mathbf{R}^3$, the *linking number* of α with β, $lk(\alpha, \beta)$, is the sum of the signed undercrossings of α with β in a regular projection of $\alpha \cup \beta$. The sign (± 1) of an undercrossing is determined by a choice of an orientation at the undercrossings.

For example, the linking of a two-component link (the Hopf link) is given in the figure with a choice of undercrossing orientations.

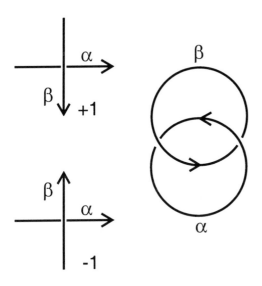

Left: Oriented undercrossings; Right: $lk(\alpha, \beta) = -1$.

Liouville's function   The arithmetic function, denoted λ, which, for any positive integer $n = p_1^{i_1} \ldots p_k^{i_k}$, returns the number $\lambda(n) = (-1)^{i_1 + \ldots + i_k}$. (*See* arithmetic function.) For example, $\lambda(540) = \lambda(2^2 \cdot 3^3 \cdot 5) = (-1)^6 = 1$. It is completely multiplicative.

locally n-connected topological space   A topological space such that, for every point p, every neighborhood of p contains a smaller neighborhood of p which is n-connected. A connected topological space X is n-connected if for every $k \leq n$, every map of the k-dimensional sphere into X is homotopic to a constant map. For example, any manifold is locally n-connected for every n, as is every locally finite simplicial complex. The one point union of infinitely many n-spheres, with the weak topology, is locally $n - 1$-connected but not locally n-connected.

logical connective   Used to build new propositional (sentential) or first-order formulas from existing ones. The usual *logical connectives* are $\wedge$ (and), $\vee$ (or), $\neg$ (not), $\rightarrow$ (implies), and $\leftrightarrow$ (if and only if). For example, if A and B are well-formed propositional formulas, then so are $(A \wedge B)$, $(A \vee B)$, $(\neg A)$, $(A \rightarrow B)$ and $(A \leftrightarrow B)$.

The truth tables for these logical connectives are as follows, where T is interpreted as true and F is interpreted as false.

A	$(\neg A)$
T	F
F	T

A	B	$(A \wedge B)$
T	T	T
T	F	F
F	T	F
F	F	F

A	B	$(A \vee B)$
T	T	T
T	F	T
F	T	T
F	F	F

A	B	$(A \rightarrow B)$
T	T	T
T	F	F
F	T	T
F	F	T

A	B	$(A \leftrightarrow B)$
T	T	T
T	F	F
F	T	F
F	F	T

logical consequence   In propositional (sentential) logic, a well-formed formula β is a *logical consequence* of a well-formed formula α if α logically implies β; i.e., if every truth assignment that satisfies α also satisfies β. For example, if A and B are sentence symbols, then A is a logical consequence of $(A \wedge B)$. In addition, β is a logical consequence of a set Γ of well-formed formulas if Γ logically implies β; i.e., if every truth assignment that satisfies every member of Γ also satisfies β.

For first order logic, let $\mathcal{L}$ be a first order language, and let α and β be well-formed formulas of $\mathcal{L}$. Then β is a logical consequence of α if α logically implies β; i.e., if, for every structure $\mathcal{A}$ for $\mathcal{L}$ and for every $s : V \to A$, whenever $\mathcal{A}$ satisfies α with s, $\mathcal{A}$ also satisfies β with s. (Here, V is the set of variables of $\mathcal{L}$ and A is the universe of $\mathcal{A}$.) In addition, β is a logical consequence of a set Γ of well-formed formulas of $\mathcal{L}$ if Γ logically implies β; i.e., if, for every structure $\mathcal{A}$ for $\mathcal{L}$ and for every $s : V \to A$, whenever $\mathcal{A}$ satisfies every member of Γ with s, $\mathcal{A}$ also satisfies β.

logically equivalent   In propositional (sentential) logic, well-formed formulas α and β are *logically equivalent* if α logically implies β and β logically implies α; that is, if every truth assignment either satisfies both α and β, or both $\neg\alpha$ and $\neg\beta$. For example, if A and B are sentence symbols, then $(\neg(A \vee B))$ and $((\neg A) \wedge (\neg B))$ are logically equivalent.

For first order logic, let $\mathcal{L}$ be a first order language, and let α and β be well-formed formulas of $\mathcal{L}$. Then α and β are logically equivalent if α logically implies β and β logically implies α; that is, if for every structure $\mathcal{A}$ for $\mathcal{L}$ and for every $s : V \to A$, $\mathcal{A}$ satisfies α with s if and only if $\mathcal{A}$ satisfies β with s. (Here, V is the set of variables of $\mathcal{L}$ and A is the universe of $\mathcal{A}$.)

logically implies   In propositional (sentential) logic, a well-formed formula α *logically implies* another well-formed formula β (notation: $\alpha \models \beta$) if every truth assignment that satisfies α also satisfies β. A set Γ of well-formed formulas logically implies a well-formed formula β (notation: $\Gamma \models \beta$) if every truth assignment that satisfies every member of Γ also satisfies β. For example, if A, B, and C are sentence symbols, $\Gamma = \{A, (A \to B)\}$, and $\beta = B$, then Γ logically implies β. This notion in propositional logic is called "tautologically implies" by some authors.

For first order logic, let $\mathcal{L}$ be a first order language, and let α and β be well-formed formulas of $\mathcal{L}$. Then α logically implies β (notation: $\alpha \models \beta$) if, for every structure $\mathcal{A}$ for $\mathcal{L}$ and for every $s : V \to A$ such that $\mathcal{A}$ satisfies α with s, $\mathcal{A}$ also satisfies β with s. (Here, V is the set of variables of $\mathcal{L}$ and A is the universe of $\mathcal{A}$.) A set Γ of well-formed formulas of $\mathcal{L}$ logically implies a well-formed formula β of $\mathcal{L}$ (notation: $\Gamma \models \beta$) if, for every structure $\mathcal{A}$ for $\mathcal{L}$ and for every $s : V \to A$ such that $\mathcal{A}$ satisfies every member of Γ with s, $\mathcal{A}$ also satisfies β with s.

Loop Theorem   A theorem addressing the conditions in which a loop in the boundary of a three-dimensional manifold which is contractible within the manifold has an equivalent embedding which bounds a disk.

Specifically, let M be a compact three-dimensional manifold and let N be a component of its boundary. If the kernel of the homomorphism $\pi_1(N) \to \pi_1(M)$ is non-trivial, then there exists a disk $D^2 \subset M$ such that $\partial D^2 \subset N$ is a simple loop that is not homotopic to zero in M.

lower bound   Let S be a subset of a partially ordered set $(P, \leq)$. An element $x \in P$ is a *lower bound* for S if $x \leq s$ for all $s \in S$.

lower limit topology   *See* Sorgenfrey line.

Luzin space   An uncountable regular topological space that has no isolated points and in which every nowhere dense set is countable. A *Luzin space* is hereditarily ccc and hereditarily Lindelöf. (*See* countable chain condition.) Because of this, Luzin spaces are related to L-spaces, which are hereditarily Lindelöf but not hereditarily separable.

M

magic square A square array of positive integers such that the sum of all its rows, columns, and diagonals are equal. Often an additional condition is added; namely, that the entries of the $n \times n$ magic square include all of the integers $1, 2, \ldots, n^2$. An example of a 3×3 magic square of this type is

$$\begin{array}{ccc} 8 & 1 & 6 \\ 3 & 5 & 7 \\ 4 & 9 & 2 \end{array}.$$

Mangoldt function The arithmetic function, denoted Λ, which is defined as follows: $\Lambda(n) = \log p$ if $n = p^i$ for some prime number p and positive integer i, and $\Lambda(n) = 0$ otherwise. (*See* arithmetic function.) For example, $\Lambda(8) = \log 2$, $\Lambda(15) = 0$. This function plays an important part in elementary proofs of the prime number theorem.

manifold A topological space M with the property that each point P possesses a neighborhood that is homeomorphic to Euclidean space $\mathbf{R}^n$, for some n. If M is connected, the dimension n is constant, and M is an n-manifold. Usually, but not always, it is desirable to assume also that M is Hausdorff and metrizable.

mapping *See* function.

mapping cylinder Given a map $f : X \to Y$ between topological spaces, the *mapping cylinder* $\mathbf{I}_f$ of f is the quotient space of the disjoint union of $X \times [0, 1]$ and Y obtained by identifying each point $(x, 0) \in X \times 0$ with $f(x) \in Y$. The space $\mathbf{I}_f$ is homotopy equivalent to Y and the map $f : X \to Y$ is homotopically equivalent to the natural inclusion $i : X \to \mathbf{I}_f$. The mapping cylinder thus justifies the statement that, in homotopy theory, "every map is equivalent to an inclusion." There is also an algebraic version of the mapping cylinder when X and Y are chain complexes.

Martin's Axiom (MA) If $\mathcal{P}$ is a partial order with the countable chain condition and $\mathcal{D}$ is a collection of fewer than continuum-many dense subsets of $\mathcal{P}$, there is a filter $G \subseteq \mathcal{P}$ which meets every $D \in \mathcal{D}$. That is, as long as $\mathcal{P}$ has no uncountable collections of incompatible elements (antichains), generic filters are able to meet any set $\mathcal{D}$ of dense subsets with $|\mathcal{D}| < 2^\omega$. In a partial order, $D \subseteq \mathcal{P}$ is dense if for any $p \in \mathcal{P}$ there is a $q \in D$ with $q \leq p$. A topological equivalent is: if $\mathcal{D}$ is a collection of fewer than continuum-many dense open sets of a compact Hausdorff space X with the countable chain condition, then $\cap \mathcal{D}$ is dense in X.

For an infinite cardinal κ, MA_κ is the statement that if $\mathcal{D}$ is a collection of dense subsets of a ccc partial order $\mathcal{P}$ with $|\mathcal{D}| \leq \kappa$, then there is a filter $G \subseteq \mathcal{P}$ such that $G \cap D \neq \emptyset$ for each $D \in \mathcal{D}$. Thus, *Martin's Axiom (MA)* is the assertion that for all $\kappa < 2^\omega$, MA_κ.

MA_ω is a theorem of ZFC, and so the Continuum Hypothesis implies MA. However, MA is also consistent with $\neg$CH. Some consequences of MA_{ω_1} are Suslin's Hypothesis (there are no Suslin lines or trees), the union of ω_1 measure zero sets has measure zero, and the union of ω_1 meager sets is meager.

mathematical induction *See* induction.

maximal element Given a set A and an ordering $\leq$ on A, $m \in A$ is said to be a *maximal element* of A if there does not exist $x \in A$ with $m < x$. Alternatively, $m \in A$ is a maximal element of A if, for all $x \in A$, if $m \leq x$, then $m = x$. Note that if A has a greatest or maximum element, then it is unique, and it is also the unique maximal element of A. If A has no greatest element, then A may have more than one, or no, maximal elements. *See* greatest element.

maximum element A greatest element of a set A with an ordering $\leq$. *See* greatest element.

mean curvature The arithmetical mean of the principal curvatures of the surface S at P. It is half the trace of the second fundamental form of the surface at P. (Some writers do not divide by 2.) More generally, the mean curvature at P in a hypersurface S of $\mathbf{R}^{n+1}$ is $\frac{1}{n}$ times the trace of the second fundamental form.

mediant   The *mediant* of two rational numbers $\frac{p}{q}$ and $\frac{r}{s}$ is the rational number $\frac{p+r}{q+s}$. For example, the mediant of $\frac{2}{3}$ and $\frac{5}{8}$ is $\frac{2+5}{3+8} = \frac{7}{11}$. The mediant of two positive rational numbers is always between the two rational numbers.

member of a set   Any object that belongs to a given set, that is, is an element of that set. For example, the number 5 is a member of the set $\{a, 5, 2\}$. Notation: $x \in S$ (x is a member of S), and $x \notin S$ (x is not a member of S).

meridian of a sphere   An inclusion of S^{n-1} in S^n which splits S^n into two equal size halves. The equator is a meridian of the sphere S^2.

Mersenne number   A number of the form $M_n = 2^n - 1$, where n is a positive integer. Determining which *Mersenne numbers* are prime has long interested mathematicians. *See also* Mersenne prime.

Mersenne prime   A number of the form $M_n = 2^n - 1$, where n is a positive integer, which is prime. For example, $M_2 = 3$ and $M_5 = 31$ are *Mersenne primes*. If M_n is a Mersenne prime, then n is prime. However, the converse is not true: $M_{11} = 23 \cdot 89$. In fact, Mersenne primes are rare. There are currently 35 known Mersenne primes, the largest being $M_{1398269}$. It is unknown whether there are infinitely many Mersenne primes.

Mersenne primes are named for Marin Mersenne, a 17th century monk who made a conjecture regarding which primes $p \leq 257$ are such that M_p is a Mersenne prime. He was later shown to have made errors of both commission and omission in his conjecture.

metric   A function $d : X \times X \rightarrow \mathbf{R}$ satisfying
(i,) $d(x, y) > 0$ if $x \neq y$; $d(x, x) = 0$,
(ii.) $d(x, y) = d(y, x)$ and
(iii.) $d(x, y) + d(y, z) \leq d(x, z)$, for all $x, y, z \in X$.

A *metric* may be interpreted as a distance function on the set X.

metric space   A topological space X equipped with a metric d such that the topology of X is that induced by d. *See* metric. Specifically, given $x \in X$ define the ε-ball about x by $B_\varepsilon(x) = \{y :$ $d(x, y) < \varepsilon\}$. Then the ε-balls $B_\varepsilon(x)$ for all $x \in X$ and $\varepsilon > 0$ form a basis for the topology of X.

metrizable space   A topological space X such that there exists a metric d on X for which the topology on X is the metric topology induced by d. *See* metric, metric space.

Meyer-Vietoris sequence   A long exact sequence in homology (or cohomology) that is obtained when a topological space X is the union of two subspaces X_1 and X_2 such that the inclusion $(X_1, X_1 \cap X_2) \rightarrow (X, X_2)$ (viewed as a map of pairs) induces an isomorphism in relative homology. The exact sequence is of the form

$$\cdots \rightarrow H_p(X_1 \cap X_2) \rightarrow H_p(X_1) \oplus H_p(X_2) \rightarrow$$
$$H_p(X) \rightarrow H_{p-1}(X_1 \cap X_2) \rightarrow \cdots$$

The *Mayer-Vietoris sequence* is closely related to the Excision Theorem for singular theory.

microbundle   A pair of maps $i : B \longrightarrow E$ and $j : E \longrightarrow B$ such that ji is the identity map on B and for each b in B, there are open neighborhoods U of b and V of ib with $iU \subset V$ and $jV \subset U$ and a homeomorphism $h : V \longrightarrow U \times R^n$ with the following properties.

(i.) The map hi restricted to U includes U as $U \times \{0\}$ in $U \times \mathbf{R}^n$.

(ii.) The map h followed by projection onto U is equal to the restriction of j to the set V.

The integer n is called the fiber dimension of the *microbundle*.

Microbundles were introduced in an attempt to construct tangent bundles on manifolds without differentiable structures. J. Milnor (Microbundles I. Topology 3 (1964) suppl. 1, 53–80) uses microbundles to show that there is a topological manifold M such that no Cartesian product $M \times M'$ has a differentiable structure that agrees with the original topological structure.

minimal element   Given a set A and an ordering $\leq$ on A, $m \in A$ is said to be a *minimal element* of A if there does not exist $x \in A$ with $x < m$. Alternatively, $m \in A$ is a minimal element of A if, for all $x \in A$, if $x \leq m$, then $m = x$. Note that if A has a least or minimum element, then it is unique, and it is also the unique minimal element of A. If A has no least element, then A may have more than one, or no, minimal elements. *See* least element.

minimal surface A surface in $\mathbf{R}^3$ with mean curvature vanishing at every point. (*See* mean curvature.) Equivalently, a *minimal surface* is a critical point for the surface area functional. This definition generalizes to surfaces in higher dimensional spaces or more general Riemannian manifolds. A subtlety of the term is the fact that a minimal surface need not minimize area; such surfaces are called *stable* minimal surfaces.

minimum element A least element of a set A with an ordering $\leq$. *See* least element.

mixed area A useful concept in convex geometry, based on the observation that one can form a weighted average of convex figures to obtain a new convex figure. If M and M_1 are convex figures in the plane and $0 \leq s \leq 1$, the mixed figure M_s is formed by taking all points $sP + tQ$ for which P is a point in M, Q is a point in M_1, and $t = 1 - s$. If $A(M_s)$ is the area of the convex figure M_s, then $A(M_s) = s^2 A(M) + 2st A(M, M_1) + t^2 A(M_1)$, where the number $A(M, M_1)$ is the *mixed area* of M and M_1.

Möbius band The rectangle $\{(x, y) \in \mathbf{R} \times \mathbf{R} : 0 \leq x, y \leq 1\}$, with the identification $(0, y) \sim (1, 1 - y)$ for $0 \leq y \leq 1$. With the usual topology, the Möbius band is a nonorientable manifold.

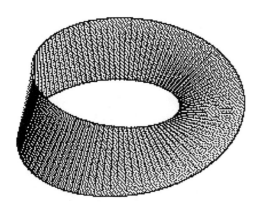

Möbius band.

Möbius function The arithmetic function, denoted μ, which is defined as follows: $\mu(1) = 1$; $\mu(n) = (-1)^k$ if n is square-free and has k distinct prime divisors; and $\mu(n) = 0$ if n is not square-free. (*See* arithmetic function.) For example, $\mu(30) = (-1)^3 = -1$, $\mu(18) = 0$. It is multiplicative.

Möbius inversion formula Let f be an arithmetic function. Define the arithmetic function F by

$$F(n) = \sum_d f(d) \cdot g\left(\frac{n}{d}\right) ,$$

where d ranges over the positive divisors of n. Then

$$f(n) = \sum_d F(d) \cdot \mu\left(\frac{n}{d}\right) ,$$

where μ is the Möbius function. In other words, F is the Dirichlet convolution of f and u (the unit function) if and only if f is the Dirichlet convolution of F and μ. *See* arithmetic function.

Möbius transformation *See* linear fractional function.

Möbius transformation group The projective linear group PL$(2, \mathbf{C})$ of all *Möbius transformations*. *See* linear fractional function. Named after the German mathematician August Ferdinand Möbius.

model Let $\mathcal{L}$ be a first order language, σ be a sentence of $\mathcal{L}$, and $\mathcal{A}$ be a structure for $\mathcal{L}$. If $\mathcal{A}$ satisfies σ with some (and hence every) $s : V \to A$, then $\mathcal{A}$ is a *model* of σ, and σ is true in $\mathcal{A}$. (Here, V is the set of variables of $\mathcal{L}$ and A is the universe of $\mathcal{A}$.) If Σ is a set of sentences of $\mathcal{L}$, then $\mathcal{A}$ is a model of Σ if $\mathcal{A}$ is a model of every sentence in Σ.

The term model is sometimes synonymous with the term structure. *See also* structure.

model complete A theory T of a first order language $\mathcal{L}$ such that, for all structures $\mathcal{A}$ and $\mathcal{B}$ which are models of T, if $\mathcal{A}$ is a substructure of $\mathcal{B}$, then $\mathcal{A}$ is an elementary substructure of $\mathcal{B}$.

As an example, let $\mathcal{L}$ be the first order language with equality, whose only predicate symbol is $<$, and let $\mathcal{R}$ be the structure for $\mathcal{L}$ whose universe is the set **R** of real numbers and where $<$ is interpreted in the usual way. Then the theory of $\mathcal{R}$, the set of all sentences true in $\mathcal{R}$, is *model complete*.

modus ponens   The logical rule of inference "from A and $(A \to B)$, infer B." Here, A and B can be any well-formed propositional (sentential) or first order formulas. Literally, *modus ponens* means "the positing method", where "to posit" means "to present as a fact" or "to postulate".

morphism   A category has objects and morphisms. Though a *morphism* is a primitive notion in category theory, it can be understood as an abstraction of the notion of function. The following categories are standard examples (the objects are listed first, the morphisms second): topological spaces and continuous functions; Abelian groups and group homomorphisms; rings and ring homomorphisms; partially ordered sets and monotone functions; complex Banach spaces and bounded linear transformations; sets and injective functions; sets and surjective functions.

motion   An element of the group of motions. *See* group of motions.

multiple   The integer c is a *multiple* of the integer a if there exists an integer b so that $ab = c$. That is, c is a multiple of a if a is a divisor of c. *See also* divisor.

multiplicative function   An arithmetic function f having the property that $f(mn) = f(m) \cdot f(n)$ whenever m and n are relatively prime. (*See* arithmetic function.) Many important functions, including the Euler phi function and the Möbius function μ, are multiplicative. The values of a *multiplicative function* depend only on its values at powers of primes: if $n = p_1^{i_1} \cdots p_k^{i_k}$ and f is multiplicative, then

$$ f(n) = f(p_1^{i_1}) \cdots f(p_k^{i_k}) . $$

See also completely multiplicative function, strongly multiplicative function.

mutually relatively prime set of integers
A set of integers such that there is no integer d greater than 1 which is a divisor of all members of the set. For instance, the set $\{2, 3, 4\}$ is *mutually relatively prime* since the only common positive divisor of 2, 3, and 4 is 1. Note that the set is not *pairwise* relatively prime since the greatest common divisor of 2 and 4 is 2. *See also* pairwise relatively prime numbers.

N

natural equivalence   *See* natural transformation.

natural isomorphism   *See* natural transformation.

natural number   A positive integer. The set of *natural numbers* is denoted **N** or $\mathbb{N}$.

natural transformation   Let C, D be categories, and let F, $G: C \to D$ be functors. A *natural transformation* is a correspondence ϕ that sends every object A of C to a morphism ϕ_A of D such that, for every morphism $f: A \to B$ of C, the diagram

$$
\begin{array}{ccc}
F(A) & \xrightarrow{F(f)} & F(B) \\
\phi_A \downarrow & & \downarrow \phi_B \\
G(A) & \xrightarrow{G(f)} & G(B)
\end{array}
$$

commutes. The correspondence ϕ is a *natural equivalence* (or *natural isomorphism*) if, in addition, ϕ_A is an isomorphism in D, for each object A of C.

negative number   A real number that is less than 0.

negative of a number   If n is a number, then *negative n* (also referred to as *the opposite of n*), is the number $-n = (-1) \times n$ (i.e., the product of -1 and n). Alternatively, $-n$ is the additive inverse of the integer n (the unique integer k so that $n + k = 0$).

neighborhood   A *neighborhood* of a point x in a topological space X is a set U such that U contains an open subset V of X with $x \in V$.

neutral geometry   The portion of geometry that can be derived without the use of Euclid's parallel postulate. This is also referred to as "absolute geometry", a term coined by Janos Bolyai.

non-Euclidean   Not satisfying the postulates from Euclid's *Elements*.

non-Euclidean geometry   A class of geometrical systems not satisfying the postulates from Euclid's *Elements*. Includes elliptic geometry, hyperbolic geometry, projective geometry, and spherical geometry.

non-Euclidean space   A space satisfying axioms that contradict the postulates from Euclid's *Elements*.

non-Euclidean surface   A surface that is a subset of a non-Euclidean space. *See* non-Euclidean space.

nonprincipal ultrafilter   An ultrafilter $\mathcal{U}$ over a Boolean algebra B with no $b \in B$ such that $\mathcal{U} = \{x \in B : b \leq x\}$.

normal bundle   When a manifold is contained in $\mathbf{R}^n$, the directions perpendicular to the tangent directions are normal. Forming a vector space at each point of the manifold, these directions yield a *normal bundle* over the manifold.

Example: Let M denote a Möbius band in $\mathbf{R}^3$. The normal bundle of M in $\mathbf{R}^3$ can be visualized by looking at the part of the bundle over the middle circle of M: this part of the bundle is again a Möbius band. One can see this by taking a normal direction (perpendicular to the surface of the Möbius band) at any point and walking around the band along the middle circle. The normal vector will be pointing in the opposite direction. This is not true for the cylinder ($S^1 \times \mathbf{R}^1$). *See* Möbius band.

normal curvature   At a point P on a surface, the curvature (with proper choice of sign) of the curve formed by the intersection of the surface with the plane through the normal vector at P and a unit vector in the tangent plane.

normalized vector   A vector made to be of length one by multiplying the vector by the reciprocal of its length.

normal space   A topological space X satisfying the following: given any two disjoint closed subsets C and D of X, there exist disjoint open

1-58488-050-3/01/$0.00+$.50
© 2001 by CRC Press LLC

subsets U and V of X such that $U \supset C$ and $V \supset D$.

normal to plane   A vector or line, passing through a given point in the plane, perpendicular to all lines in the plane passing through the point. If the plane in space is given by the equation

$$a(x - x_0) + b(y - y_0) + c(z - z_0) = 0 ,$$

then the vector (a, b, c) is *normal to the plane* at the point (x_0, y_0, z_0).

normal topological space   A topological space X in which one-point sets are closed and, given any two closed, disjoint subsets A_1, A_2 of X, there exist disjoint open subsets U_1 and U_2 of X such that $A_1 \subset U_1$ and $A_2 \subset U_2$. Examples of normal spaces include metric spaces and compact Hausdorff spaces.

normal to surface   At a point on the surface, the vector orthogonal to the tangent vector space at the point.

normal vector   A vector at a point of a manifold (contained in $\mathbf{R}^n$) that is perpendicular to all tangent vectors at that point. For example, the north pole is normal to the surface of the earth.

nowhere dense subset   A subset A of a topological space X such that the closure of A contains no nonempty open subsets of X. Any discrete set is nowhere dense in a Hausdorff space. A more interesting example is the Cantor Set, which is not discrete and yet is a *nowhere dense subset* of the unit interval $[0, 1]$. *See* Cantor set.

n-sphere bundle   A fiber bundle whose fiber is the n-dimensional sphere, and whose structure group is a subgroup of the orthogonal group $O(n + 1)$. It consists of a base space B, a total space E, and a projection map $\pi : E \longrightarrow B$. There is a covering of B by open sets U_i and homeomorphisms $\phi_i : U_i \times S^n \longrightarrow \pi^{-1}(U_i)$ such that $\pi \circ \phi_i(x, q) = x$. This identifies $\pi^{-1}(x)$ with the n-sphere. When two sets U_i and U_j overlap, the two identifications are related by orthogonal transformations $g_{ij}(x)$ of S^n. For example, if M is a surface in $\mathbf{R}^3$, then the space of vectors of length one tangent to the surface form the total space of a 1-sphere bundle.

null object   *See* zero object.

null set   *See* empty set.

number field   *See* algebraic number field.

number of distinct prime divisors function The arithmetic function, denoted ω, which, for any positive integer n, returns the number of distinct prime divisors of n. (*See* arithmetic function.) For example, $\omega(12) = \omega(24) = 2$ ($\{2, 3\}$ is the set of distinct prime divisors in both cases). It is additive.

number of divisors function   The arithmetic function, usually denoted τ or d, which, for any positive integer n, returns the number of positive divisors of n, i.e., $\tau(n) = \#\{a : 1 \le a \le n$ and $a|n \}$. (*See* arithmetic function.) For example, $\tau(12) = 6$ ($\{1, 2, 3, 4, 6, 12\}$ is the set of divisors). It is multiplicative; its value at a prime power is given by

$$\tau(p^i) = i + 1 .$$

See also sum of kth powers of divisors function.

number system   A logically organized method for expressing numbers which may be visual (using writing or hand signs), oral (spoken), or tactile (e.g., the Braille system). A variety of *number systems* have been used throughout history. The number system used by most cultures today is a positional base-10 system. *See* base of number system.

number theoretic function   *See* arithmetic function.

number theory   That branch of mathematics involving the study of the integers and their generalizations.

numeral   A physical representation of a number, often in written form.

numerator   The number a in the fraction $\frac{a}{b}$.

numerical   Of or relating to numbers or computations involving numbers.

O

object   A category has objects and morphisms. Though the notion of an *object* is a primitive in category theory, objects can be understood as generalizing or abstracting concrete mathematical entities. The following categories are standard examples (the objects are listed first, the morphisms second): topological spaces and continuous functions; Abelian groups and group homomorphisms; rings and ring homomorphisms; partially ordered sets and monotone functions; complex Banach spaces and bounded linear transformations; sets and injective functions; sets and surjective functions.

oblique   Neither perpendicular nor horizontal.

oblique angle   Any angle that is not $0°, 90°$, $180°$, or $270°$.

oblique cylinder   A cylinder that is not a right cylinder.

oblique triangle   A triangle that does not contain a right angle.

obstruction class   A cohomology class or homotopy class of maps for which being null-homologous or homotopic to zero is equivalent to the existence of the extension of some map.

Example: Suppose X is formed from a space A by attaching an n-cell D^n to A along its boundary; that is, X is the union of A and D^n with each point in S^{n-1}, the boundary of D^n, identified with some point in A by a map δ. Let $f : A \longrightarrow Y$ be a map; it will extend to a map $X \longrightarrow Y$ exactly when the class given by $f \circ \delta$ is zero in the cohomology group $H^n(X, A; \pi_{n-1}(Y))$. In particular, a map from a sphere S^{n-1} can be extended to a map from the disk D^n exactly when its class in $H^n(D^n, S^{n-1}; \pi_{n-1}(Y)) = H^n(S^n; \pi_{n-1}(Y))$ is the zero class.

obstruction cocycle   A cocycle that represents an obstruction class in cohomology. *See* obstruction class.

obtuse angle   An angle greater than $90°$ and less than $180°$.

obtuse triangle   A triangle containing an obtuse angle. *See* obtuse angle.

octagon   A polygon having eight sides.

octahedron   A polyhedron with eight faces. The regular *octahedron,* one of the five platonic solids, has 8 triagonal faces, 12 edges, and 6 vertices.

omits   A model A of a theory T *omits* a type Φ if and only if it does not realize it. That is, A omits Φ if and only if there is no n-tuple $\bar{a}$ of elements of A such that $A \models \phi(\bar{a})$ for every $\phi(\bar{x})$ in $\Phi(\bar{x})$.

one   (**1**) The smallest positive integer, denoted 1.

(**2**) The multiplicative identity of the complex numbers (and therefore of the real numbers, the rational numbers, and the integers). That is, if z is a complex number, then $1 \cdot z = z \cdot 1 = z$.

one-point compactification   A compact space X_c obtained from a given topological space X by adjoining a single point ∞ to X. The definition of the topology on X_c requires that X be a locally compact Hausdorff space. The open sets in X_c are then defined to be the open sets of X and any set of the form $V \cup \{\infty\}$ where V is an open subset of X whose complement in X is a compact set. Note that X is a subspace of X_c. The *one-point compactification* of the real line is homeomorphic to a circle, while the one-point compactification of the plane is homeomorphic to a sphere. The latter example is especially important in complex analysis where the homeomorphism is called stereographic projection. *See* stereographic projection.

one-to-one correspondence   Any function that is both one-to-one (injective) and onto (surjective); also known as a *bijective function*, or a *bijection*. For example, the function $f : \mathbf{R} \to \mathbf{R}$

1-58488-050-3/01/$0.00+$.50

given by $f(x) = 3x - 2$ is a *one-to-one correspondence*.

one-to-one function   Any function $f : A \rightarrow B$, where A and B are arbitrary sets, such that for every $x, y \in A$, $f(x) = f(y)$ implies $x = y$. Also known as an *injective function*, or an *injection*. For example, the function $f : \mathbf{N} \rightarrow \mathbf{R}$ given by $f(n) = \sqrt{n}$ is one-to-one.

onto   If A and B are arbitrary sets, any function $f : A \rightarrow B$ such that for every $y \in B$ there exists $x \in A$ satisfying $f(x) = y$ is an *onto* function. Also known as a surjective function, or as a surjection. For example, the function $f : \mathbf{R} \rightarrow \mathbf{R}$ given by $f(x) = x^3$ is onto.

open ball   In a metric space X, any set of the form $B = \{y : d(x, y) < r\}$, for some center $x \in X$ and radius $r > 0$. In a metric space, the set of *open balls* forms a basis for the metric topology. For example, in $\mathbf{R}^3$ with the usual distance metric, the open balls are just the interiors of spheres.

open cover   An *open cover* of a subspace A of a topological space X is a collection $\{U_\alpha\}$ of open subsets of X such that the union of all the U_α contains A. Open covers figure in the definition of compactness. *See* compact.

open disk   An open ball in $\mathbf{R}^2$ with the usual distance metric. *See* open ball. That is, D is an *open disk* with center $x = (x_1, x_2) \in \mathbf{R}^2$ and radius $r > 0$ if

$$D = \{y \in \mathbf{R}^2 : d(x, y) < r\}$$
$$= \{(y_1, y_2) : \sqrt{(x_1 - y_1)^2 + (x_2 - y_2)^2} < r\} .$$

open formula   A well-formed formula α of a first order language $\mathcal{L}$ such that α is quantifier-free; i.e., α does not have any quantifiers.

open map   A function $f : X \rightarrow Y$ such that the image $f(U)$ of any open set U of X is an open set in Y. If f is invertible, then f is an *open map* if and only if $f^{-1} : Y \rightarrow X$ is continuous.

open n-ball   An open ball in $\mathbf{R}^n$ with the usual distance metric. *See* open ball. That is, B is an

open n-ball with center

$$x = (x_1, x_2, \ldots, x_n) \in \mathbf{R}^n$$

and radius $r > 0$ if

$$B = \{y \in \mathbf{R}^n : d(x, y) < r\} ,$$

where

$$d(x, y) =$$
$$\sqrt{(x_1 - y_1)^2 + (x_2 - y_2)^2 + \cdots + (x_n - y_n)^2} .$$

open set   A subset U of a topological space X which belongs to the topology on X.

open simplex   The interior $\text{Int}(\sigma)$ of a simplex σ. Specifically, $\text{Int}(\sigma) = \sigma \setminus \text{Bd}(\sigma)$, where $\text{Bd}(\sigma)$, the boundary of σ, is the union of all proper faces of σ. For example, an open 1-simplex is an open interval, while an open 2-simplex is the interior of a triangle. *See* simplex.

open star   If S is a simplicial complex and v is a vertex of S, the *open star* of the vertex v is defined to be the union of the interiors of all simplices σ of S that have v as a vertex. *See* simplicial complex.

opposite angles   Two angles on a polygon (having an even number of sides) having an equal number of angles between them, regardless of the direction around which one counts.

opposite angle/side   A side and an angle on a polygon (with an odd number of sides) having an equal number of sides between them, regardless of the direction around which one counts.

opposite sides   A pair of sides on a polygon (having an even number of sides) having an equal number of sides between them, regardless of the direction around which one counts.

ordered n-tuple   A list of n arbitrary objects with a specified order, viewed as a single object. The first component of the n-tuple is the object listed first, the nth component is the object listed last, etc. For example, $(10, 10, \pi, \sqrt{2}, b)$ is an ordered 5-tuple; the third component is π.

ordered pair   An ordered list of two objects. The first (second) component of the *ordered pair* is the object listed first (second). For example, $(-1, A)$ is an ordered pair with -1 and A as its first and second components, respectively.

ordered set   *See* partially ordered set.

ordered triple   An ordered n-tuple with $n = 3$. *See* ordered n-tuple.

ordering   A partial *ordering* (on a set A). *See* partial ordering.

order topology   The topology on a set X, with a linear order relation, with a basis consisting of all intervals of the form (a, b) for any $a, b \in X$. If X has either a minimal element m or a maximal element M, then the sets $[m, b)$ and $(a, M]$ are included as well. On the real line, the *order topology* is the standard topology; that is, the topology with a basis consisting of the open intervals.

order type (of a well-ordered set)   The unique ordinal number that is order-isomorphic to the given well-ordered set. Thus, the set $\{-2, 1, 5\}$, which is well ordered by the relation $-2 < 1 < 5$, has *order type* 3. The set $\mathbf{N} \cup \{\#\}$, which is well ordered by the relation $0 < 1 < 2 < 3 < \cdots < \#$, has order type $\omega + 1$.

ordinal (or ordinal number)   A transitive set that is strictly well ordered by the element relation $\in$. For example, the *ordinal number* 3 is the set $\{0, 1, 2\}$; it is a transitive set and it is well ordered by $\in$.

ordinary helix   A curve lying on a cylinder which forms a constant angle with the elements of the cylinder.

ordinate   The y-coordinate of a point in the Cartesian xy-plane is the *ordinate* of that point. For example, the ordinate of the point $(2, -3)$ is -3.

orientable fiber bundle   A fiber bundle $F \longrightarrow E \longrightarrow B$, with F a connected compact n-manifold, such that it is possible to choose elements in the homology $H_n(F_b)$ of the fiber

above each point b in B so that around each point there is a neighborhood U and a generator of the homology $H_n(E|_U)$ of E restricted to U so that the inclusion of the fiber into $E|_U$ induces a map that takes the (chosen) generator of $H_n(F_b)$ to the (chosen) generator of $H_n(E|_U)$.

Examples: Any trivial bundle over an orientable manifold is again orientable. But the Möbius band, as a bundle over S^1, is not orientable.

orientation   A specific choice of direction for a vector space, simplex, or cell. The definitions of orientation for these objects extend to give the important notions of an orientation on a manifold, simplicial complex, or cell complex.

(1) For an n-dimensional vector space V, an *orientation* is determined by the choice of an ordered basis $\{v_1, \ldots, v_n\}$. A second basis $\{w_1, \ldots, w_n\}$ gives the same orientation precisely when the determinant of the change of basis matrix from $\{v_1, \ldots, v_n\}$ to $\{w_1, \ldots, w_n\}$ is positive.   For example, the ordered basis $\{(1, 0, 0), (0, 1, 0), (0, 0, 1)\}$ gives the orientation of $\mathbf{R}^3$ known as the *right-handed* orientation.

(2) An *orientation* of a simplex is a specific ordering of its vertices; two orderings are equivalent if one is an even permutation of the other.

(3) An *orientation* of a cell e_n (homeomorphic to an n-dimensional ball) is a choice of generator for the infinite cyclic relative homology group $H_n(e_n, \mathrm{Bd}(e_n))$.

(4) Given a manifold M, each point is contained in an open set that is homeomorphic to $\mathbf{R}^n$. Thus, using the definition of orientation for vector spaces above, M can be covered by open sets such that each open set has an *orientation*. If two such open sets U and V are not disjoint, we may ask if the orientation on $U \cap V$ inherited from U is the same or opposite to that inherited from V. If the orientation on $U \cap V$ is the same either way, we say U and V are *coherently oriented*. An orientation on a manifold M is defined to be a choice of coherently oriented open sets that cover M.

(5) Similarly, in a simplicial or cell complex, each point is contained in a simplex or cell which can be given an *orientation*. An orientation of the complex is a coherent choice of orientation for each cell or simplex.

Note that if a manifold or complex can be given one orientation, then it can also be given the reverse orientation. Many manifolds and complexes cannot be oriented. For example, an open Möbius band cannot be oriented. *See* Möbius band.

orientation preserving mapping   Any map between oriented bundles or oriented manifolds which maps the orientation of the domain to the orientation of the codomain. (Since homology is natural, a map between manifolds induces a map from the homology of one bundle or manifold to the other.)

orientation reversing mapping   Any map between oriented bundles or oriented manifolds which maps the orientation of the domain to the negative orientation of the codomain. Example: Any reflection reverses orientation: the map $S^2 \longrightarrow S^2$ given by reflection in the equator (horizontal plane) sends the generator of $H_2(S^2)$ to its negative generator.

oriented complex   A simplicial or cell complex with an orientation. *See* orientation.

orthocenter   The point of intersection of the three altitudes of a triangle.

orthogonal   At right angles.

orthogonal complement   Given a subspace W of a vector space V, the unique subspace U of V such that $V = U \oplus W$ and every vector in U is perpendicular to every vector in W.

orthogonal coordinate system   A coordinate system in which, whenever $i \neq j$, the vector with a 1 in the ith position and zeros in every other is orthogonal to the vector with a 1 in the jth position and zeros in every other.

orthogonal frame   In differential geometry, the ordered set $(x, \mathbf{v}_1, \ldots, \mathbf{v}_n)$ consisting of a point x and orthonormal vectors $\mathbf{v}_1, \ldots, \mathbf{v}_n$.

orthogonal group   The group of all $n \times n$ orthogonal matrices under multiplication. An orthogonal matrix is one whose inverse equals its transpose.

orthogonal projection   A linear transformation $T : V \to V$ from an inner product space V to itself such that $T = T^2 = T^*$, where T^* denotes the adjoint of T.

orthogonal transformation   A linear transformation whose matrix A is an orthogonal matrix, i.e., $A^{-1} = A^t$.

orthogonal vectors   A set of vectors that are pairwise orthogonal.

orthonormal   A set of vectors that are orthogonal and have magnitude 1.

orthonormalization   A process by which a set of independent vectors may be transformed into an orthonormal set of equal size while spanning the same space.

osculating circle   Given a point P on a curve, the circle that is the limit (if this exists) as a point Q approaches P along the curve of circles passing through Q and tangent to C at P.

osculating process   A method, due to P. Koebe (1912), of proving the existence of Green's function on any simply or multiply connected domain in the complex plane.

oval   Any egg-shaped curve. More generally, the boundary of a convex body in $\mathbf{R}^2$.

ovaloid   The boundary of a convex body in $\mathbf{R}^3$.

P

pair of relatively prime integers   Two integers whose greatest common divisor is 1. For example, 6 and 25 are relatively prime since gcd(6,25)= 1. *See* greatest common divisor.

pairwise disjoint   A collection of sets in which any two distinct sets are disjoint is a pairwise disjoint family. For example, $\{[2n, 2n + 1) : n \in \mathbf{N}\}$ is a collection of pairwise disjoint sets (where each interval $[2n, 2n + 1)$ is the set of real numbers x such that $2n \le x < 2n + 1$).

pairwise relatively prime numbers   A set of integers with the property that no two share a common divisor greater than 1.

parabola   The set of points in the plane equidistant from a given point and a given line. Alternatively, a conic section formed by the intersection of a circular cone with a plane such that the intersection is connected but unbounded.

paraboloid   The surface given by the set of solutions in $\mathbf{R}^3$ to an equation of the form $\frac{x^2}{a^2} + \frac{y^2}{b^2} = z$ (elliptic paraboloid) or $\frac{x^2}{a^2} - \frac{y^2}{b^2} = z$ (hyperbolic paraboloid).

paracompact topological space   A Hausdorff space with the property that each open cover has a locally finite open refinement that covers X. *See* refinement of a cover. Metric spaces are paracompact but, of course, not generally compact. Paracompactness is important in the theory of manifolds because it is a sufficient condition on a space to construct a partition of unity.

parallel   Equidistant, in some sense. In Euclidean space, two lines are *parallel* if they do not intersect and there is a plane in which they both lie.

parallelepiped   A polyhedron whose faces are parallelograms.

parallelizable manifold   A manifold whose tangent bundle is trivial.

parallelogram   A four-sided polygon having opposite sides parallel.

Parallel Postulate   The fifth postulate of Euclid's *Elements*, which requires that, if two lines are cut by a third, and if the sum of the interior angles on one side of the third is less than 180°, then the two lines will meet on that side of the third.

partially ordered set   A set with a partial ordering. *See also* partial ordering. A partially ordered set is sometimes called a *poset*.

partial ordering   A binary relation on a set A (i.e., a subset of $A \times A$), often denoted by $\le$, which is reflexive (for all $x \in A$, $x \le x$), antisymmetric (for all $x, y \in A$, if $x \le y$ and $y \le x$, then $x = y$), and transitive (for all $x, y, z \in A$, if $x \le y$ and $y \le z$, then $x \le z$). Given a *partial ordering* $\le$ on A and $x, y \in A$, $x < y$ is defined to mean $x \le y$ and $x \ne y$. Note that if $\le$ is a partial ordering on A, then it is not necessarily the case that every two distinct elements of A are comparable; i.e., there may exist $x, y \in A$ with $x \ne y$ and $x \not\le y$ and $y \not\le x$.

Sometimes one sees an alternative definition of partial ordering, where $<$ is defined first and $\le$ is defined in terms of $<$.

A partial ordering on a set A is a binary relation on A, often denoted by $<$, which is antireflexive (for all $x \in A$, $x \not< x$) and transitive. Given a partial ordering $<$ on A and $x, y \in A$, $x \le y$ is defined to mean $x < y$ or $x = y$. Using this definition of partial ordering, one can prove that $\le$ is antisymmetric. Note that if $<$ is a partial ordering on A, then it is not necessarily the case that every pair of distinct elements are comparable; i.e., there may exist $x, y \in A$ with $x \ne y$ and $x \not< y$ and $y \not< x$.

Regardless of the choice of defining $<$ from $\le$ or $\le$ from $<$, the respective notions of $\le$ and $<$ are the same.

One example of a partial ordering is the usual ordering $\le$ on the natural numbers $\mathbf{N}$. This partial ordering is in fact a total, or linear ordering. *See* linear ordering. Let $\mathcal{P}(\mathbf{N})$ be the power set of the set of natural numbers; that is, the set of all subsets of **N**. Then set containment $\subseteq$ is a partial ordering on $\mathcal{P}(\mathbf{N})$ which is not a linear ordering, as $\{1, 2\}$ and $\{3\}$ are not $\subseteq$-comparable; i.e., $\{1, 2\} \not\subseteq \{3\}$ and $\{3\} \not\subseteq \{1, 2\}$.

partial recursive function   All functions mentioned are functions on the natural numbers **N**; an n-ary function is *partial* if its domain is some subset of $\mathbf{N}^n$ (i.e., the function may not be defined on all inputs). The notion of a *partial recursive function* is a formalization (Kleene, 1936) of the notion of an intuitively computable partial function. An n-ary partial function φ is partial recursive if it can be derived from a certain set of initial functions by finitely many applications of composition, recursion, or the μ-operator; i.e., there is a finite sequence

$$\varphi_0, \varphi_1, \ldots, \varphi_k = \varphi$$

of functions such that for all $i, 0 \leq i \leq k$,
  (i.) φ_i is an initial function or
  (ii.) φ_i can be obtained from $\{\varphi_j : 0 \leq j < i\}$ by composition, recursion, or the μ-operator. The following functions are initial functions.

- $S(x) = x + 1$ (the successor function)

- $C_i^n(x_1, \ldots, x_n) = i$, for all natural numbers $i, n \geq 0$ (the constant functions)

- $P_i^n(x_1, \ldots, x_n) = x_i$, for all natural numbers $n \geq 1$ and $1 \leq i \leq n$ (the projection functions).

Let $g_1, \ldots, g_k, f$ be n-ary functions and let h be a k-ary function; let $\overline{x}$ denote an n-tuple $x_1, \ldots, x_n$. The function f is obtained from $g_1, \ldots, g_k$ and h by composition if for all natural numbers $x_1, \ldots, x_n$, $f(\overline{x}) = h(g_1(\overline{x}), \ldots, g_k(\overline{x}))$.

Let f be an n-ary function, $n \geq 1$, g be an $(n-1)$-ary function, h be an $(n+1)$-ary function, and $\overline{y}$ denote the $(n-1)$-tuple $y_1, \ldots, y_{n-1}$. The function f is obtained from g and h by recursion if for all natural numbers $x, y_1, \ldots, y_{n-1}$,

$$f(0, \overline{y}) = g(\overline{y})$$
$$f(x + 1, \overline{y}) = h(x, f(x, \overline{y}), \overline{y}).$$

Let f be an n-ary function and let g be an $(n + 1)$-ary (possibly partial) function. Let $\overline{x}$ denote an n-tuple $x_1, \ldots, x_n$, and let μ be the least number operator; i.e., $(\mu x)[\cdots]$ denotes the least natural number x satisfying property $[\cdots]$. Let $\downarrow$ be an abbreviation of the phrase "is defined". If for all natural numbers $x_1, \ldots, x_n$,

$$\begin{aligned}
f(\overline{x}) &= (\mu y)[g(\overline{x}, y) \\
&= 0 \wedge (\forall z < y)[g(\overline{x}, z) \downarrow]],
\end{aligned}$$

then f is obtained from g by the μ-operator.

If an n-ary function is partial recursive and total (i.e., the domain of the function is all of $\mathbf{N}^n$), then the function is called recursive (or total recursive).

Any primitive recursive function (*see* primitive recursive function) is recursive; for example, the function $f(x_1, x_2) = x_1 + x_2$ is recursive. However, it is not the case that every (total) recursive function is primitive recursive. An example of a recursive function that is not primitive recursive is Ackermann's function, which is defined informally by "double recursion", as follows.

$$\begin{aligned}
A(0, y) &= S(y) \\
A(x + 1, 0) &= A(x, 1) \\
A(x + 1, y + 1) &= A(x, A(x + 1, y)).
\end{aligned}$$

Ackermann's function grows faster than any primitive recursive function.

By the Church-Turing Thesis, any intuitively computable partial function (*see* Church-Turing Thesis, computable) is partial recursive. The function

$$\psi(e) = \begin{cases} 1 & \text{if} \quad \varphi_e(e) \text{ is defined} \\ \text{undefined} & \text{if} \quad \varphi_e(e) \text{ is undefined,} \end{cases}$$

where φ_e is the partial recursive function with Gödel number e, is partial recursive.

partition   (1) Of a set. A pairwise disjoint collection of nonempty subsets of the given set, whose union is the given set. For example, $\{\{3, a\}, \{-2\}\}$ is a *partition* of the set $\{3, a, -2\}$. Also, $\{[n, n + 1) : n \in \mathbf{Z}\}$ is a partition of **R** (where each interval $[n, n + 1)$ is the set of real numbers x such that $n \leq x < n + 1$). *See* quotient set.

(2) Of a positive integer. If n is a positive integer, a *partition of* n is a sequence $(k_1, k_2, \ldots, k_r)$ of positive integers such that

$k_1 \geq k_2 \geq \cdots \geq k_r$ and $k_1 + k_2 + \cdots + k_r = n$. For example, $(4, 3)$ and $(2, 2, 2, 1)$ are two partitions of 7.

Pascal's triangle   A specific triangular array of numbers, named after the 19th century French mathematician/philosopher Blaise Pascal, the first few rows of which are given below:

$$
\begin{array}{ccccccccc}
 & & & & 1 & & & & \\
 & & & 1 & & 1 & & & \\
 & & 1 & & 2 & & 1 & & \\
 & 1 & & 3 & & 3 & & 1 & \\
1 & & 4 & & 6 & & 4 & & 1
\end{array}
$$

If the first row is labeled the "0th row" (for example, the 4th row is 1 4 6 4 1) and call the first entry (on the left) of each row the "0th" entry (so the 6 in the middle of the 4th row is the 2nd entry of that row), then the kth entry of the nth row is the binomial coefficient $\binom{n}{k}$.

Although this array is known as *Pascal's triangle*, it has been found in Chinese manuscripts that were printed 500 years before Pascal's birth.

path   A *path* from a point x to a point y in a topological space X is any continuous function $f : [0, 1] \to X$ with $f(0) = x$ and $f(1) = y$. Intuitively, the path is the image of the function f.

Peano space   A compact, connected locally connected metric space. By the famous Hahn-Mazurkiewicz Theorem, Peano sets are precisely those sets that occur as continuous images of the unit interval.

Peano's Postulates   A set of axioms for developing the properties (using naive set theory) of the natural numbers. The axioms were published by Peano in 1889 and were based on work by Dedekind.

There are five Peano Postulates.

(i.) 0 is a natural number.

(ii.) For every natural number n, there is a natural number n', called the successor of n.

(iii.) For every natural number n, $n' \neq 0$.

(iv.) For any natural numbers m and n, if $m' = n'$, then $m = n$.

(v.) If I is any subset of the natural numbers such that

(a.) $0 \in I$ and

(b.)   for any natural number n, if $n \in I$, then $n' \in I$,

then I contains all natural numbers.

This last postulate is the Principle of Mathematical Induction and is applicable to each of the uncountably many subsets of the natural numbers.

Peano's Postulates uniquely determine the set of natural numbers in the sense that if M is a set that satisfies the five postulates above (with the phrase "n is a natural number" replaced by "$n \in M$"), then M is the set of natural numbers.

pedal triangle   The triangle within a given triangle formed by connecting the non-vertex endpoints of the altitudes of the given triangle.

Pell's equation   The equation $x^2 - dy^2 = k$, where x and y are unknown variables and d and k are integers. *Pell's equation* is an example of a Diophantine equation (an equation for which one searches for integer or rational solutions). The integer d is usually assumed to be square-free (that is, if p is a prime divisor of d, then p^2 is *not* a divisor of d) and positive because otherwise the equation has only finitely many integer solutions for x and y, if any.

pencil of circles   The collection of all circles in a plane passing through two given points.

pencil of lines   (1) The collection of all lines in a given plane which pass through a given point.

(2) The collection of all lines parallel to a given line.

pencil of planes   The collection of all planes in space containing a given line.

pencil of spheres   The collection of all spheres containing a given circle.

pentadecagon   A polygon having 15 sides.

pentagon   A polygon having five sides.

pentagonal number   Any positive integer of the form $\frac{3n^2 - n}{2}$ (i.e., any entry in the sequence, $1, 5, 12, 22, 35, \ldots$).

pentahedron   A polyhedron with five faces.

percent   From the Latin for "per hundred", percentages are used as alternatives to fractions or decimals to represent ratios of numerical values. For example, 50% (read "fifty percent") represents $\frac{50}{100}$ or .50. Thus, 35% of 250 is $.35 \times 250 = 87.5$ (since $\frac{35}{100} = \frac{87.5}{250}$).

perfect number   A positive integer n having the property that the sum of its positive divisors is $2n$, i.e., $\sigma(n) = 2n$. Thus, 6 is a *perfect number* since $1 + 2 + 3 + 6 = 2(6)$. The next two perfect numbers are 28 and 496. All even perfect numbers have been characterized as being of the form $2^{p-1}(2^p - 1)$, where both p and $2^p - 1$ are prime (i.e., where $2^p - 1$ is a Mersenne prime). There are no known odd perfect numbers. *See also* abundant number, deficient number, Mersenne prime.

perfect set   A topological space X with the property that every point of X is an accumulation point of X. That is, given any $x \in X$ and any neighborhood U of x, the set $(U \cap X)\backslash\{x\}$ is nonempty. All intervals on the real line are perfect. An example of a perfect subset of the real line which is not an interval is furnished by the Cantor set. *See* Cantor set.

perfect square   An integer a for which there exists another integer b such that $a = b^2$. For example, $36 = 6^2$ and $289 = 17^2$ are *perfect squares*.

perigon   A 360° angle.

perimeter   (1) The length of a closed curve, especially when considered as the boundary of a plane figure.

(2) The closed curve forming the boundary of a plane figure.

periodic continued fraction   A continued fraction

$$a_0 + \cfrac{b_1}{a_1 + \cfrac{b_2}{a_2 + \cfrac{b_3}{a_3 + \cfrac{b_4}{a_4 + \ddots}}}}$$

for which there exist positive integers p and N so that for all $k \geq N$, $a_{k+p} = a_k$ and $b_{k+p} = b_k$.

perpendicular   (1) The relative position of a pair of lines that intersect so that they form a pair of equal adjacent angles.

(2) The relative positon of a line and a plane such that the line is *perpendicular* to every line with which it intersects in the plane.

(3) The relative position of a pair of planes that intersect so that a line in one is *perpendicular* to both the line of intersection and to a line in the other which is perpendicular to the line of intersection.

perpendicular bisector   A line or line segment that bisects a given line segment and is perpendicular to it. *See* perpendicular.

pi   A number, denoted π, equal to the ratio between the circumference and diameter of any circle. It is also the ratio of the area of a circle to the square of its radius. It is known that π is a transcendental (and therefore irrational) number. The value of π is approximately 3.14159265358979 and, although the value of *every* digit in the decimal expansion of π is not known, in 1989 David and Gregory Chudnovsky calculated π to 1,011,196,691 decimal place accuracy.

point at infinity   A point of the hyperplane at infinity. *See* hyperplane at infinity.

polar coordinates   Two numbers (r, θ) that determine a point P in the (x, y)-plane, r being the distance from the origin O and $0 \leq \theta < 2\pi$ the angle that the ray OP makes with the positive x-axis, measured in radians and counterclockwise, except if $P = O$, which is associated to the number zero only; thus $x = r\cos\theta$, $y = r\sin\theta$. If r is allowed to be negative, the pair $(-r, \theta)$ gives the same point as the pair $(r, \theta + \pi)$.

Polish space   A topological space X that is separable and completely metrizable. That is, X has a countable dense subset, and there is a metric d inducing the topology on X for which every Cauchy sequence converges. *Polish spaces* are the natural setting for descriptive set theory. Examples of Polish spaces include $\mathbf{R}$, $\mathbf{R}^n$, $\mathbf{C}$, $\mathbf{C}^n$, the Cantor space $2^{\mathbf{N}}$, and the Baire space $\mathbf{N}^{\mathbf{N}}$.

Pollard rho method   A method for factoring an integer which is known to be composite (using one of the pseudoprime tests, for example). The method is based in part on the fact that, although it is difficult to find factors to large numbers, it is relatively easy (using Euclid's algorithm) to find the greatest common divisor of two integers.

Let m be a positive composite integer and define the sequence $\{u_i\}$ recursively as follows:

(i.) $u_0 = 1$;

(ii.) u_{i+1} is the unique integer so that $u_{i+1} \equiv u_i^2 + 1 (\mathrm{mod}\, m)$ and $0 \leq u_{i+1} < m$.

(In general, u_0 could be any positive integer and $u_{i+1} = f(u_i)$ for some nonlinear polynomial, f). Next, compute $D_n = \gcd(u_{2n} - u_n, m)$ for $n = 1, 2, 3, \ldots$ If $D_n \neq 1$ for any n, then D_n is a divisor of m.

To illustrate the method, let $m = 1771$. Then the first few terms of the sequence (starting with u_1) defined by $u_0 = 1$ and $u_{i+1} \equiv u_i^2 + 1 (\mathrm{mod}\, m)$ are $2, 5, 26, 677, 1412, \ldots$ Notice that $\gcd(u_2 - u_1, 1771) = \gcd(3, 1771) = 1$, and $\gcd(u_4 - u_2, 1771) = \gcd(672, 1771) = 7$, so 7 is a divisor of 1771 (as is $1771 \div 7 = 253$). Since 7 is prime and 253 is composite, we could repeat Pollard's method to show that 253 is the product of the primes 11 and 23, thus completing the factorization of 1771.

The *Pollard rho method* was introduced by J. M. Pollard in 1975.

polygonal number   A positive integer n so that n dots can be arranged in a specified polygonal pattern. Also called a figurate number. *See also* triangular number, pentagonal number.

polynomial   A finite sum of multiples of non-negative integer powers of an indeterminate x, with coefficients in a given set R. For example, $3x^5 - 4x^3 + x^2 + 7x - 12$ is a *polynomial* with coefficients in $\mathbf{Z}$.

poset   *See* partially ordered set.

positive number   A real number that is greater than 0.

positive orthant   The set of points $(x_1, x_2, \ldots, x_n)$ in $\mathbf{R}^n$ such that $x_i \geq 0$, $1 \leq i \leq n$. In the case of the plane, this is the positive quadrant.

postulates of Euclid   Euclid based his geometry on five basic assumptions, known as The Postulates. The first four postulates assert that one can join any two points by a straight line, extend straight lines continuously, and draw a circle with any given center and radius, and that all right angles are equal. The fifth postulate asserts the uniqueness of a straight line, through a point, parallel to a given line.

power set (of a set)   The set of all subsets of a given set. The power set of S is denoted by $\mathcal{P}(S)$ or 2^S. For example, if $S = \{3, 5\}$ then $\mathcal{P}(S) = \{\emptyset, \{3\}, \{5\}, \{3, 5\}\}$. The power set of any set S, together with the operations of union, intersection, and complementation with respect to S, forms a Boolean algebra. The unit and zero element of this Boolean algebra are S and $\emptyset$, respectively. *See also* Cantor's Theorem.

predicate calculus   The syntactical part of first order logic. A *predicate calculus* is a formal system consisting of a first order language, the set of all well-formed formulas, a particular set Λ of well-formed formulas, which are called logical axioms, and a list of rules of deduction.

The well-formed formulas that are logical axioms should be valid formulas. A typical axiom that might occur in a predicate calculus is

$$\forall x(\alpha \to \beta) \to (\forall x\alpha \to \forall x\beta),$$

where α and β are any well-formed formulas, or, if the language contains equality,

$$x = y \to (\alpha \to \alpha'),$$

where α is an atomic formula, and α' is obtained from α by replacing the variable x in α in zero or more places by the variable y. A typical rule of deduction in a predicate calculus is *modus ponens*. *See* modus ponens.

A predicate calculus is used to prove theorems. *See also* proof, theorem. While the actual choice of logical axioms and rules of deduction is not important, it is important that a predicate calculus be both sound (i.e., any well-formed formula which is provable in the formal system should be a logical consequence of the logical axioms) and complete (i.e., any logical consequence of the logical axioms should be provable in the formal system).

Predicate calculus is sometimes called first order predicate calculus.

predicate logic   *See* first order logic.

primary cohomology operation   A natural transformation of functors

$$H^i(X, A; M) \longrightarrow H^{i+j}(X, A; N) ;$$

an operation may be defined for many choices of i and j and many choices of Abelian groups M and N, or only for specific choices. Operations are often additive. The squaring operation, which takes u to u^2, is not additive; the Steenrod square operations (also called reduced squares) are additive.

Cohomology operations also exist on generalized cohomology theories, for example K-theory and cobordism theories. Adams operations on K-theory are cohomology operations. *See also* secondary cohomology operation.

prime factor   A prime p that is a divisor of an integer n. For example, the *prime factors* of 24 are 2 and 3. *See* divisor.

prime ideal   Let S be a set. An ideal I on S is a *prime ideal* if, for all $X \subseteq S$, either $X \in I$ or $S \setminus X \in I$.

prime number   (1) An integer with exactly two positive integer factors (including itself and 1). For example, 5 is prime because its positive integer factors are 1 and 5, while 6 is not prime because the positive integer factors of 6 are 1, 2, 3, and 6. Note that the integer 1 is not prime since it has only one positive integer factor, itself.

(2) More generally, an element p of a ring is prime (or irreducible) if it is not a unit and all of its factors (in the ring) are associates (unit multiples) of p.

Prime Number Theorem   If $\pi(x)$ denotes the number of prime numbers less than or equal to the positive real number x, then

$$\lim_{x \to \infty} \frac{\pi(x)}{\left(\frac{x}{\log(x)}\right)} = 1 .$$

That is, if x is large, $\pi(x) \approx \frac{x}{\log(x)}$. An equivalent formulation of the theorem is that $\lim_{x \to \infty} \frac{\pi(x)}{\mathrm{Li}(x)} = 1$, where $\mathrm{Li}(x) = \int_2^x \frac{dx}{\log(x)}$ (the so-called logarithmic integral). The *Prime Number Theorem* was first proved, independently, by Jacques Hadamard and Charles de la Valée Poussin in 1896.

primitive recursive function   All functions mentioned are functions on the natural numbers. An n-ary function f is *primitive recursive* if it can be derived from a certain set of initial functions by finitely many applications of composition and recursion; i.e., there is a finite sequence

$$f_0, f_1, \ldots, f_k = f$$

of functions such that for all $i, 0 \leq i \leq k$,

(i.) f_i is an initial function or
(ii.) f_i can be obtained from $\{f_j : 0 \leq j < i\}$ by composition or recursion.

The following functions are initial functions.

- $S(x) = x + 1$ (the successor function)

- $C_i^n(x_1, \ldots, x_n) = i$, for all natural numbers $i, n \geq 0$ (the constant functions)

- $P_i^n(x_1, \ldots, x_n) = x_i$, for all natural numbers $n \geq 1$ and $1 \leq i \leq n$ (the projection functions).

Let $g_1, \ldots, g_k, f$ be n-ary functions and let h be a k-ary function; let $\overline{x}$ denote an n-tuple $x_1, \ldots, x_n$. The function f is obtained from $g_1, \ldots, g_k$ and h by composition if for all natural numbers $x_1, \ldots, x_n$, $f(\overline{x}) = h(g_1(\overline{x}), \ldots, g_k(\overline{x}))$.

Let f be an n-ary function, $n \geq 1$, g be an $(n-1)$-ary function, h be an $(n+1)$-ary function, and $\overline{y}$ denote the $(n-1)$-tuple $y_1, \ldots, y_{n-1}$. The function f is obtained from g and h by recursion if for all natural numbers $x, y_1, \ldots, y_{n-1}$,

$$\begin{aligned} f(0, \overline{y}) &= g(\overline{y}) \\ f(x+1, \overline{y}) &= h(x, f(x, \overline{y}), \overline{y}). \end{aligned}$$

For example, the function $f(x, y) = x + y$ is primitive recursive. Informally, the recursion equations for f are

$$f(0, y) = y$$
$$f(x + 1, y) = S(f(x, y)).$$

More formally,

$$f(0, y) = P_1^1(y)$$
$$f(x + 1, y) = h(x, f(x, y), y),$$

where $h(x, y, z) = S(P_2^3(x, y, z))$.

principal curvature   If P is a point in a surface S in $\mathbf{R}^3$, then the *principal curvatures* at P are the minimum and maximum values of the curvatures of the curves formed by intersecting S with a plane through P containing the normal vector to the surface at P. Equivalently, the principal curvatures are the eigenvalues of the Weingarten map at P.

principal fiber bundle   A fiber bundle whose fiber is a topological group G and whose structure group is also G, acting on itself by (left) multiplication. *See* fiber bundle. It consists of a base space B, a total space E, and a projection map $\pi : E \longrightarrow B$. There is a covering of B by open sets U_i and homeomorphisms $\phi_i : U_i \times G \longrightarrow \pi^{-1}(U_i)$ such that $\pi \circ \phi_i(x, q) = x$. This identifies $\pi^{-1}(x)$ with G as a topological space. Examples of *principal fiber bundles* are constructed by taking the quotient map $\pi : L \longrightarrow (L/G)$ from a Lie group L to the quotient space of L by a closed subgroup G. A universal covering map $\pi : E \longrightarrow B$ is a principal bundle with the fundamental group of B (with the discrete topology) as fiber and group.

principal ideal   Let S be a nonempty set and let $\mathcal{P}(S)$ be the power set of S. An ideal I on S is a *principal ideal* if there exists a set $A \subseteq S$ such that $I = \{X \in \mathcal{P}(S) : X \subseteq A\}$.

principal type   A type $\Phi(\bar{x})$ of a theory T in a language L such that there is an L-formula $\theta(\bar{x})$ in $\Phi(\bar{x})$ such that $T \vdash \forall \bar{x}(\theta(\bar{x}) \rightarrow \phi(\bar{x}))$ for every $\phi(\bar{x}) \in \Phi(\bar{x})$. That is, under T, the single formula θ generates the entire set Φ.

principal ultrafilter   An ultrafilter $\mathcal{U}$ over a Boolean algebra B such that there is a $b \in B$ such that $\mathcal{U} = \{x \in B : b \leq x\}$.

principle of dependent choices   Suppose R is a binary relation on a nonempty set S, and that, for every $x \in S$, there exists $y \in S$ such that $(x, y) \in R$. Then there exists a countable sequence $x_0, x_1, \ldots, x_n, \ldots$ ($n \in \mathbf{N}$) of elements in S such that $(x_n, x_{n+1}) \in R$, for all $n \in \mathbf{N}$. This principle is also known as the Axiom of Dependent Choice. It is a consequence of, but is weaker than, the Axiom of Choice, and it is the usual replacement of the Axiom of Choice if the Axiom of Determinacy is assumed.

principle of inclusion-exclusion   A combinatorial formula for the cardinality of the union of a finite collection of finite sets. In the case of two sets, the formula is $|A \cup B| = |A| + |B| - |A \cap B|$. For r arbitrary finite sets $A_1, \ldots, A_r$, the formula is

$$|A_1 \cup \cdots \cup A_r| =$$
$$\sum_{n=1}^{r} (-1)^{n+1} \sum |A_{k_1} \cap \cdots \cap A_{k_n}|,$$

where the second sum ranges over all n-tuples of natural numbers $(k_1, \ldots, k_n)$ such that $1 \leq k_1 < \cdots < k_n \leq r$.

product   The general term used for the result obtained by applying some operation, usually called multiplication. For example, *product* of natural numbers, product of complex numbers, product of real-valued functions, Cartesian product of sets, product of matrices, product of cardinal numbers, product of ordinal numbers, product of elements of a group, product of objects in a category.

product bundle   Formed by taking the tensor product of the fibers (of two vector bundles E and E' over B) over each point of B. Thus, the tensor product of two line bundles is again a line bundle. Line bundles over a space form a group with respect to this product; the group identity is the trivial bundle $B \times R \longrightarrow B$.

product category   Let $\mathcal{C}_1, \ldots, \mathcal{C}_n$ be categories. The product $\mathcal{C}_1 \times \cdots \times \mathcal{C}_n$ is the category whose objects are n-tuples $(A_1, \ldots, A_n)$, where each A_i is an object of $\mathcal{C}_i$, and the morphisms are n-tuples $(f_1, \ldots, f_n)$, with each f_i a morphism of C_i; morphism composition is defined componentwise: $(f_1, \ldots, f_n) \circ (g_1, \ldots, g_n) = (f_1 \circ g_1, \ldots, f_n \circ g_n)$. The product of an arbitrary number of categories is defined similarly.

product metric   The metric on a finite product of metric spaces, defined by the formula

$$d((x_1, \ldots, x_n), (y_1, \ldots, y_n))$$
$$= \sum_{i=1}^{n} \frac{d_i(x_i, y_i)}{2^i},$$

where d_i is a metric on X_i and $(x_1, \ldots, x_n)$, $(y_1, \ldots, y_n) \in X_1 \times \cdots \times X_n$. See metric space. This definition shows that a finite product of metrizable spaces is metrizable.

product of cardinal numbers   The product of some cardinal numbers is the cardinal number that is equinumerous with the Cartesian product of the given cardinal numbers. If κ and λ are cardinal numbers, $k \cdot \lambda$ denotes their cardinal product. For example, $3 \cdot 2 = 6$ and $\aleph_3 \cdot \aleph_{17} = \aleph_{17}$.

product of objects   Suppose C is a category and $\{A_i : i \in I\}$ is a family of objects of C, where I is some index set. Let $p_i : A \to A_i$ be a morphism for each $i \in I$. The tuple $(A; p_i : i \in I)$ is the product of this family if, for every object B of C and every set of morphisms $f_i : B \to A_i$, for $i \in I$, there is a unique morphism $f : B \to A$ such that $p_i \circ f = f_i$, for all $i \in I$. The morphisms p_i are usually called *projection morphisms*.

product space   The Cartesian product of an arbitrary collection of topological spaces $\{X_\alpha\}_{\alpha \in A}$, with the product topology. See product topology.

product topology   The standard topology for the product $\prod_\alpha X_\alpha$ of topological spaces. A basis is given by sets of the form $\prod_\alpha U_\alpha$ where U_α is an open subset of X_α and $U_\alpha = X_\alpha$ for all but finitely many indices α. The *product topology* is the coarsest topology on the product space which makes all the projection maps continuous. See projection map.

projection map   The maps p_α from the Cartesian product $\prod_{\alpha \in A} X_\alpha$ of topological spaces,

into X_α, defined, for each α, by $p_\alpha(\{x_\alpha\}) = x_\alpha$. See also product space, product topology.

projective geometry   An axiomatic system that grew out of perspective drawing during the Renaissance; one characterization axiom says that any two lines in the projective plane have exactly one point in common. Analytically, the projective space $\mathbf{P}^n$ of dimension n can be given projective coordinates $(x_0 : \ldots : x_n)$ which determine a point up to rescaling $(\lambda x_0 : \ldots : \lambda x_n)$ by a non-zero number λ, and satisfy the condition that at least one x_i is non-zero. Thus, the set of lines containing the origin in $\mathbf{R}^3$ gives a model for the projective plane $\mathbf{P}^2$. In it, lines correspond to planes through the origin and the point of intersection to the line common to two planes.

projective set   The *projective sets* form a hierarchy extending the Borel hierarchy in any Polish space X. Let Σ_1^1 denote the collection of all analytic sets in X. For $n \geq 1$, let

$$\Pi_n^1 = \{A \subseteq X : X \setminus A \in \Sigma_n^1\}$$

and then let Σ_{n+1}^1 be the collection of all projections of Π_n^1 sets in $X \times \mathbf{N}^\mathbf{N}$, where $\mathbf{N}^\mathbf{N}$ is the Baire space. Let $\Delta_n^1 = \Sigma_n^1 \cap \Pi_n^1$. Then the sets in

$$\mathbf{P} = \bigcup_n \Sigma_n^1 = \bigcup_n \Pi_n^1$$

are the projective sets. They form a hierarchy because for each $n \geq 1$,

$$\Sigma_n^1 \cup \Pi_n^1 \subseteq \Delta_{n+1}^1 = \Sigma_{n+1}^1 \cap \Pi_{n+1}^1 .$$

In addition,

$$\Delta_1^1 = \text{Borel} .$$

The projective classes Σ_n^1, Π_n^1, and Δ_n^1 are also known as Lusin point classes.

proof   In first order logic, let $\mathcal{L}$ be a first order language and consider a particular predicate calculus for $\mathcal{L}$, with Λ the set of logical axioms. Let α be a well-formed formula of $\mathcal{L}$. A *proof* of α in the predicate calculus is a sequence $\alpha_1, \alpha_2, \ldots, \alpha_n$ of well-formed formulas of $\mathcal{L}$ such that $\alpha_n = \alpha$ and such that for all i, $1 \leq i \leq n$, either

    (i.) $\alpha_i \in \Lambda$ (i.e., α_i is a logical axiom) or (ii.) there exist $1 \leq j_1 < \cdots < j_k < i$ such that α_i can be deduced from $\alpha_{j_1}, \ldots, \alpha_{j_k}$ using a rule of inference. (The value of k depends on the rule of inference; for example, if the rule of inference is modus ponens, then $k = 2$.)

The formula α is provable from the predicate calculus (notation: $\vdash \alpha$) if there is a proof of α from the predicate calculus.

If Γ is a set of well-formed formulas of $\mathcal{L}$, then a proof of α from Γ is a sequence $\alpha_1, \alpha_2, \ldots, \alpha_n$ of well-formed formulas of $\mathcal{L}$ such that $\alpha_n = \alpha$ and such that for all i, $1 \leq i \leq n$, either

(i.) $\alpha_i \in \Gamma \cup \Lambda$ or

(ii.) there exist $1 \leq j_1 < \cdots < j_k < i$ such that α_i can be deduced from $\alpha_{j_1}, \ldots, \alpha_{j_k}$ using a rule of inference.

The formula α is provable from Γ in the predicate calculus (notation: $\Gamma \vdash \alpha$) if there is a proof of α from Γ in the predicate calculus.

The notion of proof in propositional logic is entirely analogous.

The notion of proof in formal logic is also called formal proof or deduction.

proper fraction   A positive rational number $\frac{a}{b}$ where a and b are positive integers and $a < b$. For example, $\frac{15}{38}$ is a *proper fraction,* while $\frac{15}{8}$ is an *improper* fraction.

properly discontinuous transformation group
A group G acts properly discontinuously on a space X if, for each x in X, there is an open neighborhood U so that whenever g is not the identity in G, $U \cap gU$ is empty.

Properly discontinuous transformation groups are useful for studying covering spaces, which are given by maps $X \longrightarrow Y$ such that any point in Y has a (connected) neighborhood U whose inverse image is the disjoint union of open sets of X each homeomorphic to U. A particular instance of this is the covering of a homogeneous space by a Lie group: an inclusion of Lie groups $H \longrightarrow G$ has coset space G/H which inherits the quotient topology from G; the quotient map $G \longrightarrow G/H$ is a covering map.

proper subset (of a set)   A set S is a *proper subset* of a set T if S is a subset of T but S is not equal to T. Thus, a set is never a proper subset of itself. For example, $\{3, 10\}$ is a proper subset of $\{3, 10, 47\}$. The notation is not entirely uniform.

If S is a proper subset of T, many authors denote this by $S \subset T$, whereas others choose to denote it by $S \subsetneq T$ or $S \subsetneq T$, for example. *Compare with* subset.

propositional calculus   The syntactical part of propositional logic. A *propositional calculus* is a formal system, consisting of an alphabet (*see* propositional logic), the set of all well-formed formulas, a particular set of well-formed formulas, which are called axioms, and a list of rules of deduction.

The well-formed formulas that are axioms are formulas that are intuitively obvious, and should be tautologies. A typical axiom that might occur in a propositional calculus would be

$$(\alpha \to (\beta \to \alpha)),$$

where α and β are any well-formed propositional formulas (this is actually called an axiom scheme, rather than an axiom, since there are infinitely many axioms of this form, one for each different choice of α and β). A typical rule of deduction in a propositional calculus is *modus ponens*. *See* modus ponens.

A propositional calculus, using the axioms and rules of deduction, is used to prove theorems. *See also* proof, theorem. While the actual choice of axioms and rules of deduction is not important, it is important that a propositional calculus be both sound (i.e., any well-formed formula that can be proved from the formal system should be a tautology) and complete (i.e., any tautology should be provable from the formal system).

Propositional calculus is also called sentential, or statement calculus.

propositional logic   A formal logic with the following alphabet of symbols:

(i.) $($, $)$ (parentheses)

(ii.) $\neg, \vee, \wedge, \to, \leftrightarrow$ (logical connectives)

(iii.) $A_1, A_2, A_3, \ldots$ (nonlogical symbols).

Often the list of logical connectives in item (ii.) is shortened to some complete list of logical connectives, such as $\{\neg, \to\}$. The symbol A_n is called the nth propositional symbol, or nth sentential or sentence symbol.

The propositional symbols have no meaning, although using truth assignments they can be interpreted as either true or false. *Propositional logic* has rules that tell which expressions from the language are well-formed formulas. The propositional calculus (sentential calculus) is used to produce theorems of propositional logic. *See* propositional calculus, proof. Truth assignments lead to a semantic notion of truth in propositional logic, while the propositional calculus gives a syntactical notion of provability. Propositional logic is also called sentential logic.

pseudocompact topological space   A topological space X with the property that every real-valued continuous function defined on X is bounded. Pseudocompact spaces play a significant role in the theory of C^*-algebras.

pseudomanifold   A simplicial complex S which is a union of n-simplices (for some n) and satisfies

(i.) each $n - 1$-simplex of S is the face of exactly two n-simplices, and

(ii.) given two n-simplices σ and σ' there exists a sequence of n-simplices $\sigma = \sigma_0, \sigma_1,$ $\ldots, \sigma_k = \sigma'$ with $\sigma_i \cap \sigma_{i+1}$ an $n - 1$ simplex of S.

Psuedomanifolds share a key homological property with actual manifolds; namely, $H_n(S) = \mathbf{Z}$ if S is orientable and $H_n(S) = 0$ otherwise.

pseudoprime   (**1**) An odd composite integer, n, with the property that $2^n \equiv 2 \pmod{n}$. That is, n is a *pseudoprime* if n is a divisor of $2^n - 2$.

The name is derived from the fact that if p is a prime number, then $a^p \equiv a \pmod{p}$ for all integers a.

(**2**) A composite integer n so that $a^n \equiv a$ $\pmod{n}$ for all integers a is an *absolute pseudoprime*.

Pythagorean field   A field, F, in which the sum of the squares of any two elements from the field is the square of an element from the field. That is, F is Pythagorean if, for every a and b in F, there exists a c in F so that $a^2 + b^2 = c^2$. The rational numbers are not Pythagorean since $1^2 + 1^2 = 2$ is not the square of a rational number. However, the real numbers are a *Pythagorean field* since $\sqrt{a^2 + b^2}$ is a real number whenever a and b are real.

Pythagorean triple   A triple of positive integers (a, b, c) satisfying the equation $a^2 + b^2 = c^2$. For example, $(6, 8, 10)$ is a *Pythagorean triple* since $6^2 + 8^2 = 10^2$. If (a, b, c) is a Pythagorean triple and a, b, and c are pairwise relatively prime, then (a, b, c) is known as a *primitive* Pythagorean triple $((3, 4, 5)$ and $(5, 12, 13)$ are primitive Pythagorean triples).

It can be shown that either a or b (or both) must be even if (a, b, c) is a Pythagorean triple. In fact, (a, b, c) is a Pythagorean triple *if and only if* there exist positive integers k, m, and n so that $\gcd(m, n) = 1$, exactly one of m or n is even, $a = (m^2 - n^2)k$, $b = 2mnk$, and $c = (m^2 + n^2)k$, providing a formula for generating all Pythagorean triples.

Q

quantifier   *Quantifiers* are used in order to quantify if elements with a certain property exist in a particular universe. The quantifiers are denoted symbolically by $\exists$ (the *existential quantifier*) and $\forall$ (the *universal quantifier*). The interpretation of the existential quantifier $(\exists x)[\dots]$ is that there exists an object x (possibly more than one) in the universe with property $[\dots]$. The interpretation of the universal quantifier $(\forall x)[\dots]$ is that all objects x in the universe have property $[\dots]$.

Note that only one quantifier suffices, since $(\forall x)[\dots]$ is logically equivalent to $\neg(\exists x)\neg[\dots]$.

A typical use of quantifiers occurs among the axioms for group theory. Given a set G with a binary operation $\cdot$ on G (i.e., a function from $G \times G$ to G),

$$(\exists e)(\forall x)[x \cdot e = e \cdot x = x]$$

is the axiom that states that there is an element of G that functions as an identity, while

$$(\forall x)(\exists x^*)[x \cdot x^* = x^* \cdot x = e]$$

is the axiom that states that every element of G has an inverse in G. Here, the quantifiers range over the universe (group) G.

Also note that order of quantifiers is essential. The statement $(\forall x)(\exists y)[\dots]$ does not necessarily have the same meaning as $(\exists y)(\forall x)[\dots]$. In the first statement, the y that exists may depend on the choice of x, while in the second statement, the y that exists does not depend on x.

quantifier elimination   Let $\mathcal{L}$ be a first order language. A theory T of $\mathcal{L}$ admits *quantifier elimination* if, for every well-formed formula φ of $\mathcal{L}$, there is a quantifier-free formula ψ of $\mathcal{L}$ such that $(\varphi \leftrightarrow \psi)$ is a theorem of T ($T \vdash (\varphi \leftrightarrow \psi)$).

quasicomponent   Given a topological space X, define an equivalence relation $\sim$ on X by setting $x \sim y$ if there are no open sets U and V of X with $x \in U$, $y \in V$, and $X = U \cup V$. The *quasicomponent* of x in X is the equivalence class of x under this equivalence relation.

quotient category   If $\mathcal{C}$ is a category and $\sim$ is a congruence relation on $\mathcal{C}$, the *quotient category* of $\mathcal{C}$ with respect to $\sim$ is the category $\mathcal{C}'$ whose objects are the objects of $\mathcal{C}$ and whose morphisms are the equivalence classes of morphisms of $\mathcal{C}$ (with respect to $\sim$); morphism composition $\bar{\circ}$ in $\mathcal{C}'$ is given by $\overline{f_1} \,\bar{\circ}\, \overline{f_2} = \overline{f_1 \circ f_2}$, where $\overline{f_i}$ is the equivalence class of f_i with respect to $\sim$.

quotient map   A surjective function $p : X \to Y$ between topological spaces such that a subset U is open in Y if and only if $p^{-1}(U)$ is open in X.

If $\sim$ is an equivalence relation on X, then the map $q : X \to X/\sim$, sending points of X to their equivalence classes, is sometimes referred to as a *quotient map*. This agrees with the present definition for the equivalence relation given by $x \sim y$ if and only if $p(x) = p(y)$.

quotient object   If A is an object of a category $\mathcal{C}$, a *quotient object* A is an ordered pair (f, A'), where A' is an object of $\mathcal{C}$ and f is an epimorphism $f : A \to A'$. For example, in the category of groups and group homomorphisms, a quotient object of the additive group $\mathbf{Z}$ is the pair $(f, \mathbf{Z}_2)$, where f is the group homomorphism that sends even integers to 0 and odd integers to 1. The dual notion of a quotient object is the notion of a subobject. *See* subobject.

quotient set   A set that is a partition of some other set. In practice, a *quotient set* is obtained as the collection of equivalence classes of an equivalence relation on some set. For example, the set

$$\{\{\text{all even integers}\}, \{\text{all odd integers}\}\}$$

is a quotient set since it is a partition of $\mathbf{Z}$. The equivalence relation on $\mathbf{Z}$ that gives rise to this quotient set is: $a \sim b$ if and only if $a - b$ is even.

quotient space   Let X be a topological space and $\sim$ an equivalence relation on X. Let X^* be the set of distinct equivalence classes $[x]$ of X.

1-58488-050-3/01/$0.00+$.50
© 2001 by CRC Press LLC

Define $p : X \rightarrow X^*$ by $p(x) = [x]$ and give the set X^* the quotient topology corresponding to the surjective map p. *See* quotient topology. Then X^* is called a *quotient space* of X. Notice that equivalent points of X are identified to a single point in X^*. This construction thus gives a topological method for factoring out subspaces of a space X analogous to the quotient group construction in group theory.

quotient topology   Given a surjective function $p : X \rightarrow Y$ where X is a topological space and Y is a set, the *quotient topology* on Y induced by p is the topology that makes p a quotient map. (*See* quotient map.) That is, a subset U of Y is defined to be an open set of Y if and only if $p^{-1}(U)$ is an open subset of the topological space X.

R

radical   *See* root of a number.

radius of regular polygon   The radius of the unique circle passing through all the vertices of a regular polygon.

radius of regular polyhedron   The radius of the unique sphere passing through all the vertices of a regular polyhedron.

radix   A synonym for base. *See* base of number system. For example, the decimal expansion of a real number is also known as its base 10 expansion or its expansion to the *radix* 10.

range   (1) *Range* of a function. The set of all values attained by a function. The range of a function f is often denoted by ran(f). Thus, if $f: A \to B$ is a function, ran(f) = $\{y \in B : (\exists x \in A)\, f(x) = y\}$. For example, if $f: \mathbf{R} \to \mathbf{R}$ is the function given by $f(x) = x^2$, the range of f is $[0, \infty)$. *Compare with* image.

(2) *Range* of a variable. The set of all values a given variable can attain.

(3) *Range* of a binary relation. If R is a binary relation, the range of R, often denoted by ran(R), is the set $\{x : (\exists y)\, (x, y) \in R\}$.

rank   The *rank* of a set x is the least ordinal α such that $x \in V_{\alpha+1}$. In particular, for any ordinal α, rank(α) = α.

The same notion can be defined using ϵ-recursion:

$$\text{rank}(x) = \sup\{\text{rank}(y) + 1 : y \in x\},$$

assuming sup$\{\emptyset\} = 0$.

rational function   A function that is expressible as a quotient of polynomials.

rational number   A real number that can be expressed as a quotient of integers. Furthermore, the digits of the decimal expansion of a real number will consist of a sequence which,

eventually, repeats periodically if and only if the number is a *rational number*. The set of all rational numbers is normally denoted $\mathbf{Q}$ or $\mathbb{Q}$.

real analytic fiber bundle   A fiber bundle such that the base space B is a real analytic manifold, the group G is a Lie group, and the coordinate transformations $g_{ij} : U_i \cap U_j \longrightarrow G$ are real analytic maps. *See* fiber bundle.

realizes   A model A of a theory T *realizes* a type Φ if there is a set of elements in A which satisfies every formula in Φ. More precisely, A realizes $\Phi(\bar{x})$, where $\bar{x} = \{x_1, \dots, x_n\}$ contains all free variables occurring in the formulas in Φ, if and only if there is an n-tuple $\bar{a}$ of elements of A such that $A \models \phi(\bar{a})$ for every $\phi(\bar{x})$ in $\Phi(\bar{x})$.

real number   A number that has an infinite decimal representation, which may or may not terminate or repeat. If the decimal representation repeats or terminates, the *real number* is a rational number; otherwise it is irrational.

The set of real numbers (usually denoted $\mathbf{R}$ or $\mathbb{R}$) can be constructed as the *completion* of the rational numbers, in the sense that every real number is the limit of an infinite sequence of (not necessarily distinct) rational numbers.

reciprocal   If r is a nonzero real number, then its *reciprocal* is the number $\frac{1}{r}$. For example, the reciprocal of $\frac{2}{3}$ is $\frac{3}{2}$ and the reciprocal of $\sqrt{2}$ is $\frac{1}{\sqrt{2}} = \frac{\sqrt{2}}{2}$.

recursion   Let f be an n-ary function ($n \geq 1$), g be an $(n-1)$-ary function, h be an $(n+1)$-ary function, and $\bar{y}$ denote the $(n-1)$-tuple $y_1, \dots, y_{n-1}$ (all functions are functions on the natural numbers). The function f is obtained from g and h by *recursion* if, for all natural numbers $y_1, \dots, y_{n-1}$,

$$\begin{aligned} f(0, \bar{y}) &= g(\bar{y}) \\ f(x+1, \bar{y}) &= h(x, f(x, \bar{y}), \bar{y}). \end{aligned}$$

The function $f(x, y) = x + y$ can be defined by recursion as follows. Let $S(x)$ be the successor function; i.e., $S(x) = x + 1$ for all x. Informally, the recursion equations for f are

$$\begin{aligned} f(0, y) &= y \\ f(x+1, y) &= S(f(x, y)). \end{aligned}$$

1-58488-050-3/01/$0.00+$.50
© 2001 by CRC Press LLC

More formally, f is obtained from g and h by recursion as

$$f(0, y) = g(y)$$
$$f(x + 1, y) = h(x, f(x, y), y),$$

where g is the identity function $g(y) = y$ for all y, $h(x, y, z) = S(P_2^3(x, y, z))$, and $P_2^3(x, y, z) = y$ for all x, y, z.

Similarly, the factorial function $p(n) = n!$ for all n is obtained by recursion informally as

$$0! = 1$$
$$(n + 1)! = (n + 1) \cdot n!.$$

recursive   A function f on the natural numbers which is total (i.e., the domain of f is the set of all natural numbers) and partial recursive. *See* partial recursive function. By the Church-Turing Thesis, the phrase "f is recursive" can mean f is total and computable in any formal sense. *See* Church-Turing Thesis, computable.

A set A of natural numbers is *recursive* if its characteristic function is recursive; i.e., the function

$$\chi_A(n) = \begin{cases} 1 & \text{if } n \in A \\ 0 & \text{if } n \notin A \end{cases}$$

is recursive.

The set $A = \{n \in \mathbf{N} : n \text{ is prime}\}$ is recursive.

recursively enumerable   A set A of natural numbers which is the domain of some partial recursive function. Equivalently, A is *recursively enumerable* if it is the empty set or it is the range of some (total) recursive function; i.e., if A is non-empty, then there is a computable function $f : \mathbf{N} \to \mathbf{N}$ which "lists" the elements of A. If A is the domain of the partial recursive function with Gödel number e, then A is denoted by W_e, the eth recursively enumerable set.

For example, the halting set

$$K = \{e : \varphi_e(e) \text{ is defined}\}$$

(the set of all numbers e such that the partial recursive function with Gödel number e on input e is defined) is recursively enumerable, but not recursive. The set

$$\{e : (\forall x)[\varphi_e(x) \text{ is defined}]\}$$

is not recursively enumerable.

The term *computably enumerable* (c.e.) is synonymous with recursively enumerable.

reductio ad absurdum   Literally, "a leading back to the nonsensical". In mathematics, *reductio ad absurdum* means proof by contradiction. In a proof by contradiction, one assumes the hypotheses of the statement to be proved, as well as the negation of the statement to be proved. The proof is complete when a contradiction (i.e., a situation where both a statement A and its negation $\neg A$ are true) is encountered, and one then concludes the statement which was to be proved.

reduction of a language   Let $\mathcal{L}_1$ and $\mathcal{L}_2$ be first order languages. The language $\mathcal{L}_1$ is a *reduction* of $\mathcal{L}_2$ if $\mathcal{L}_1 \subseteq \mathcal{L}_2$; i.e., if $\mathcal{L}_2$ is an expansion of $\mathcal{L}_1$. *See* expansion of a language.

reduct of a structure   Let $\mathcal{L}_1$ and $\mathcal{L}_2$ be first order languages such that $\mathcal{L}_1$ is a reduction of $\mathcal{L}_2$. Let $\mathcal{A}$ be a structure for $\mathcal{L}_2$. The *reduct* of $\mathcal{A}$ to $\mathcal{L}_1$ is the structure obtained from $\mathcal{A}$ which gives only the interpretations of the predicate, constant, and function symbols in $\mathcal{L}_1$ (all interpretations of the additional symbols in $\mathcal{L}_2$ are discarded). *See* expansion of a structure.

Reeb foliation   A *Reeb foliation* of the sphere S^3 is a codimension-one foliation in which one leaf is a torus $S^1 \times S^1$, dividing the sphere into two solid tori, and each remaining leaf diffeomorphic to the plane $\mathbf{R}^2$. Thus, the sphere is represented as a union of surfaces, only one of which is compact, such that at each point there is a local coordinate system in which each plane $\{z = constant\}$ is contained in a single surface.

Reeb Stability Theorem   Let M be a smooth manifold with a codimension-one foliation in which one of the leaves is diffeomorphic to the sphere S^2. Then all leaves of the foliation must be spheres or projective planes and the manifold M must be $S^2 \times S^1$ or the connected sum of two copies of real projective space $\mathbf{RP}^3$.

refinement of a cover   Given a cover $\{A_\alpha\}$ of a topological space X (that is, $X \subset \cup A_\alpha$), a second cover $\{B_\beta\}$ is *a refinement of* $\{A_\alpha\}$ if each B_β is contained in some A_α.

reflexive relation   A binary relation R on some set S such that $(x, x) \in R$, for every $x \in S$. For example, $\leq$ on $\mathbf{Z}$, the usual order relation on the set of integers, is a *reflexive relation.*

regular cardinal   A cardinal κ such that the cofinality of κ is κ. Thus, κ is regular if, for any increasing sequence of ordinals $\gamma_\alpha < \kappa$ whose length is less than κ, sup γ_α is also less than κ. For example, any successor cardinal is regular (assuming the axiom of choice), whereas $\aleph_\omega$ is not.

regular covering   A concept arising in the theory of covering spaces. Assume that B is an arcwise connected, locally arcwise connected space. A continuous function $\pi : E \longrightarrow B$ is a covering map if each point in B has an arcwise connected neighborhood U such that each component of $\pi^{-1}(U)$ is open in E and is mapped homeomorphically onto U by π. If $\gamma : [0, 1] \longrightarrow B$ is a curve, and if $\pi(e) = \gamma(0)$, then there is a unique curve $\gamma' : [0, 1] \longrightarrow E$ with $\gamma'(0) = e$ and $\pi(\gamma'(t)) = \gamma(t)$. γ' is called a lift of γ. The covering is *regular* if whenever γ is a closed curve in B, then either every lift of γ to a curve in E is closed or no lift of γ is closed. For a *regular covering,* the covering transformations are in 1-1 correspondence with $\pi^{-1}(b)$, for b any point in B.

regularly homotopic immersions   An immersion is a differentiable map $f : M \longrightarrow N$, whose derivative $df(m)$ is nonsingular at every point p. It is locally an embedding, but it need not be globally 1-1. Two such immersions, f_0 and f_1, are *regularly homotopic* if there is a differentiable map $H : M \times [0, 1] \longrightarrow N$ with $H(m, 0) = f_0(m)$ and $H(m, 1) = f_1(m)$ for all m in M, such that if f_t is defined by $f_t(m) = H(m, t)$, then f_t is an immersion for each t in $[0, 1]$. Thus, the immersion f_0 can be continuously deformed through immersions to f_1 in such a way that tangent vectors to M vary continuously through the deformation. This rules out untying a knot by pulling it tight.

regular neighborhood   A *regular neighborhood* N of a subcomplex L in a simplicial complex K is a neighborhood of L that is formed from the simplices of the second derived subdivision of K.

If $\{\sigma''\}$ is the collection of simplices of the second derived subdivision of K, then the regular neighborhood of L, denoted $N(L)$ is the collection of simplexes

$$N(L) = \{\sigma'' : \sigma'' \cap L \neq \emptyset\}.$$

regular polygon   A (convex) polygon whose sides have equal length, in which case the vertices lie on a circle. The n vertices of a regular n-gon can be taken to be the points $(\cos \frac{2\pi i}{n}, \sin \frac{2\pi i}{n})$, $i = 0, \ldots, n$.

regular polyhedron   A (convex) polyhedron all of whose faces are congruent and all of whose vertices belong to the same number of faces. There exist only five types, the Platonic solids: tetrahedron, hexahedron, octahedron, dodecahedron, icosahedron (resp. 4,6,8,12,20 faces).

regular topological space   A topological space X in which one-point sets are closed and, given any closed subset A of X and a point $x \in X$ not in A, there exist disjoint open subsets U and V of X such that $A \subset U$ and $x \in V$.

relation   (1) n-ary relation. A set of n-tuples.
(2) Binary relation. A set of ordered pairs.
(3) Ternary relation. A set of ordered triples.
(4) Binary relation on a set. If S is an arbitrary set, R is a *binary relation* on S if R is a subset of $S \times S$. For example, the relation "m divides n" is a binary relation on the set of natural numbers, since it is the set $\{(m, n) : m$ divides $n\}$ which is a subset of $\mathbf{N} \times \mathbf{N}$.

relative complement of a set (with respect to another set)   The *relative complement of a set* S, with respect to a set U, denoted by $U \backslash S$ or $U - S$, is the set of elements that are in U but not in S. For example, the relative complement of the set of even integers with respect to $\mathbf{Z}$ is the set of odd integers.

relative computability   Let $\mathbf{N}$ be the set of natural numbers, φ be a (possibly partial) function on $\mathbf{N}$ and f be a total function on $\mathbf{N}$. (A function is partial if its domain is some subset of $\mathbf{N}$; a function is total if its domain is all of $\mathbf{N}$.) Intuitively, the function φ is *computable relative to* f if there is an algorithm, or effective procedure, which, given $n \in \mathbf{N}$ as input, produces $\varphi(n)$ as output, in finitely many steps, if $n \in \text{dom}(\varphi)$, where the algorithm is allowed to make finitely many queries about values $f(y_1), \ldots, f(y_k)$ of f. Similarly, a set A of natural numbers is computable relative to a set B of natural numbers if there is some algorithm which, given $n \in \mathbf{N}$ as input, outputs "yes" if $n \in A$ or "no" if $n \notin A$ after finitely many steps, where the algorithm is allowed to make finitely many queries about membership $y_1 \in B?, \ldots, y_k \in B?$ in B.

To formalize the notion of *relative computability* with a mathematical definition, one relativizes one of the formal definitions of computability (*see* computable, partial recursive function). For example, let φ be a partial function on $\mathbf{N}$ and let A be a set of natural numbers. The function φ is partial recursive in A (or partial recursive, relative to A) if it can be derived from the initial functions, together with χ_A (the characteristic function of A), by finitely many applications of composition, recursion, or the μ-operator. A set B of natural numbers is recursive in A if its characteristic function χ_B is partial recursive in A. *See* partial recursive function. Similarly, the function φ is Turing computable in A (or Turing reducible to A, or Turing computable, relative to A) if there is some oracle Turing machine which computes it, using an oracle for A. A set B is Turing computable in A if its characteristic function χ_B is Turing computable in A.

An oracle Turing machine is a Turing machine with a bi-infinite, read-only oracle tape, which is separate from its work tape (where the input is originally written), on which is written the characteristic function χ_A of A; i.e., the tape looks like

$$\ldots B \; n_0 \; n_1 \; n_2 \; \ldots,$$

where for all $i \geq 0$, $n_i = 1$ if $i \in A$, $n_i = 0$ if $i \notin A$, and $\ldots B$ indicates that all cells to the left of the value of $\chi_A(0)$ are blank. The operation of the Turing machine is the same, except that the transition function is modified to account for the character being scanned on the oracle tape. In other words, if Q is the set of machine states, S is the work tape alphabet, and $O = \{B, 0, 1\}$ is the oracle tape alphabet, then the transition function δ has domain some subset of $Q \times S \times O$ and range contained in $Q \times S \times \{R, L\}$, and where $\delta(q, a, b) = (\hat{q}, \hat{a}, m)$ means that if the machine is in state q, scanning symbol a on its work tape and b on its oracle tape, then it replaces a by $\hat{a}$ on its work tape, moves the read head on both tapes one cell to the left or right, depending on the value of m, and goes into state $\hat{q}$. Computation of a partial function is the same as for a non-oracle Turing machine; at the beginning of the computation, the read head on the oracle tape is positioned on the cell giving the value of $\chi_A(0)$. *See* computable for a description of a Turing machine.

By a relativized version of the Church-Turing Thesis, all formalizations of relative computability capture the intuitive notion of relative computability, and so the phrase φ is computable, relative to A (or φ is recursive in A, or φ is A-computable, or φ is A-recursive) means that φ is computable relative to A in any such formalization. The notation $\varphi \leq_T A$ means that φ is computable, relative to A. The relation $\leq_T$ is called Turing reducibility.

As an example, note that for any set of natural numbers A, its complement $\overline{A}$ in $\mathbf{N}$ is computable, relative to A. In addition, any recursively enumerable (computably enumerable) set B is computable relative to the halting set $K = \{e : \varphi_e(e) \text{ is defined}\}$, where φ_e denotes the partial recursive function with Gödel number e.

relative homology group   Given a topological space X and a subspace A, the nth *relative singular homology group* $H_n(X, A)$ is the nth homology group of the chain complex $\{S_q(X)/S_q(A)\}$ obtained by taking the group $S_q(X)$ of chains on X modulo the group $S_q(A)$ of chains on the subspace A. If X is a simplicial (or cellular complex) and A is a subcomplex, the nth simplicial (or cellular) relative homology group $H_n(X, A)$ is just the ordinary homology of the complex X/A with the subcomplex A identified to a point. A long exact sequence of the form

$$\cdots \to H_n(A) \to \to H_n(X) \to H_n(X, A)$$
$$\to H_{n-1}(A) \to \cdots$$

relates the relative to the ordinary homology groups.

relative homotopy group   Given a topological space X and a subspace A, the nth *relative homotopy group* $\pi_n(X, A)$ is the set of homotopy equivalence classes of maps f from the n-dimensional ball B^n to X such that $f(S^{n-1}) \subseteq A$ where S^{n-1}, the $n - 1$-sphere, is the boundary of B^n. A homotopy between two such maps f and g is required to carry S^{n-1} into A for all t. If $A \subseteq B \subseteq X$, then there is a long exact sequence of the form

$$\cdots \to \pi_n(X, B) \to \to \pi_n(X, A) \to \pi_n(B, A)$$
$$\to \pi_{n-1}(X, B) \to \cdots .$$

relatively open set   A subset U of a topological space X such that U is a proper subset of a subspace A of X and U is an open subset of A.

relative topology   The topology on a subset A of a topological space X that is the collection of all intersections of A with open sets in X. That is, $U \subseteq A$ is open in the *relative topology* on A if there is an open set $V \subseteq X$ with $U = V \cap A$. The relative topology makes any set $A \subseteq X$ into a subspace of X.

repeating decimal   A decimal representation

$$\ldots a_4 a_3 a_2 a_1 a_0 . a_{-1} a_{-2} a_{-3} \ldots$$

of a real number for which there exist positive integers P and N, so that for every $n \geq N$, $a_{-n} = a_{-n-P}$, where a_{-n} (resp. a_{-n-P}) is the digit in the 10^{-n} (resp. 10^{-n-P}) place in the decimal expansion of the number. For example, the decimal expansion of the number $\frac{1517}{990}$ is $1.532323\ldots$, where the sequence 32 repeats forever (this is often written $\frac{1517}{990} = 1.5\overline{32}$, where the bar over 32 means that this sequence repeats forever). It can be shown that a real number has either a repeating or terminating decimal representation if and only if that number is a rational number. Note that a terminating decimal representation could be said to have "repeating zeros" and therefore be a *repeating decimal* representation as well.

representative of an equivalence class   Any element of an equivalence class.

restriction of a function   If $f : A \to B$ is a function, the *restriction* of f to a set S (usually S is a subset of the domain), denoted by $f|S$, is $\{(x, y) : y = f(x) \text{ and } x \in S\}$.

retract   A subspace A of a topological space X such that there exists a retraction $r : X \to A$. (*See* retraction.) *Retracts* are important in algebraic topology because they cause the long exact homology and cohomology sequences to split as short exact sequences.

retraction   A continuous function $r : X \to A$ from a topological space X to a subspace A is a *retraction* of A if $r(a) = a$ for all $a \in A$. Equivalently, r is a left inverse to the inclusion $A \hookrightarrow X$.

Riemann Hypothesis   The conjecture that all of the nontrivial zeros of the Riemann zeta function have real part $\frac{1}{2}$ (i.e., are of the form $x + iy$ where $x = \frac{1}{2}$). Bernhard Riemann stated in his memoirs that it seemed likely to be true and if proved could likely be used to prove that there are infinitely many twin prime pairs. David Hilbert listed it as one of the most important outstanding problems facing mathematicians at the dawn of the 20th century. Although it is known that there are infinitely many zeros of the zeta function with real part $\frac{1}{2}$, it is still an open problem as this book is printed at the dawn of the 21st century. *See also* Riemann zeta function, generalized Riemann hypothesis.

Riemannian geometry   The study of the geometric properties of locally Euclidean manifolds. A Riemannian manifold is a manifold whose tangent space at each point p possesses a positive definite inner product $g(p)(X, Y)$, which varies continuously (usually smoothly) with the point p. This structure allows one to define lengths, angles, and other geometric quantities. The term *Riemannian geometry* is sometimes used to refer specifically to elliptic geometry, which is a non-Euclidean geometry in which the parallel postulate is replaced by the postulate that straight lines always intersect.

Riemann zeta function   The Dirichlet series $\zeta(s) = \sum\limits_{n=1}^{\infty} \frac{1}{n^s}$ defined for (extendible to) all complex numbers $s \neq 1$. It can be shown that $\zeta(-2n) = 0$ for all positive integers n. It is conjectured that the only other zeros are of the form $s = \frac{1}{2} + iy$ (that is, have real part equal to $\frac{1}{2}$). This conjecture is known as the Riemann Hypothesis.

right adjoint of a functor   Let $\mathcal{C}$ and $\mathcal{D}$ be categories, and let $F: \mathcal{C} \to \mathcal{D}$, $G: \mathcal{D} \to \mathcal{C}$ be functors. G is the *right adjoint* of F (and F is the *left adjoint* of G) if there is a bijection θ between the collection of morphisms from $F(A)$ to B and the collection of morphisms from A to $G(B)$ that is natural for objects A of $\mathcal{C}$ and objects B of $\mathcal{D}$. Hence, this bijection θ sends every morphism $f: F(A) \to B$ to a morphism $\theta(f): A \to G(B)$ so that both conditions (i.) $\theta(f \circ F(g)) = (\theta(f)) \circ g$, and (ii.) $\theta(h \circ f) = (G(h)) \circ (\theta(f))$ are satisfied, for every pair of morphisms $g: A' \to A$ and $f: B \to B'$. For example, if $\mathcal{C}$ is the category of groups and group homomorphisms, $\mathcal{D}$ is the category of sets and functions, the forgetful functor $G: \mathcal{C} \to \mathcal{D}$ is the right adjoint to the free group functor $F: \mathcal{D} \to \mathcal{C}$.

rigid motion   An even transformation of (Euclidean) space which preserves lengths and angles. A *rigid motion* takes any geometric figure to one which is congruent to itself. Because a symmetry of the plane reverses orientation, it is not considered a rigid motion of the plane, although it is so viewed when the plane is thought of as a subset of three-dimensional space. The rigid motions form a group under composition; it is the component of the identity in the group of isometries.

root of a number   (1) If n is a positive integer and a is a complex number, an nth root of a is a complex number r such that $r^n = a$.

(2) If n is a positive integer and r is a real number, the nth root of a, denoted $\sqrt[n]{a}$, is the unique real number r so that $r^n = a$, if such a number exists. If n is odd, $\sqrt[n]{a}$ always exists (and is positive when a is positive and negative when a is negative), while if n is even and a is negative, $\sqrt[n]{a}$ does not exist, within the real numbers.

root of equation   A number that, when substituted in a given equation, makes the equation valid. For example, the real numbers $\sqrt{2}$ and $-\sqrt{2}$ are *roots of the equation* $x^2 = 2$, since $x = \pm\sqrt{2}$ are solutions to the equation. A root of a polynomial $p(x)$ is a root of the equation $p(x) = 0$.

rotation   A rigid motion of the plane which fixes exactly one point, or a rigid motion of three-dimensional space which fixes the points on exactly one line, which is the axis of *rotation*.

ruled surface   An algebraic surface, birational to $C \times \mathbf{P}^1$, where C is a smooth projective curve. For example, a smooth quadric in $\mathbf{P}^3$, which is isomorphic to $\mathbf{P}^1 \times \mathbf{P}^1$.

Russell's Paradox   A paradox of naive (informal) or non-axiomatic set theory. In naive set theory, it is possible to form the set $A = \{x : x \notin x\}$. Note that if $A \in A$, then $A \notin A$, and if $A \notin A$, then $A \in A$. This contradiction is called *Russell's Paradox*.

Russell's Paradox was discovered by Bertrand Russell in 1901 and published by him in 1903. The discovery of this and other paradoxes revealed that set theory could not be used as a language to formalize mathematics in a naive fashion, so that an axiomatic approach, giving rules for which sets could exist, needed to be developed in order to avoid contradictions. There are several axiomatizations of set theory, including ZF (Zermelo-Fraenkel set theory) and NBG (von Neumann-Bernays-Gödel set theory).

Prior to the discovery of Russell's Paradox, it was believed that any definable collection; i.e., any collection $\{x : P(x)\}$ of objects x satisfying a property $P(x)$, is a set. The difficulty with this (and with Russell's Paradox) is that some collections are, in some sense, "too big" to be sets. In ZF set theory, only definable collections that are already subsets of existing sets can be sets (this is the Axiom of Comprehension, or Axiom of Subsets). In this set theory, the collection A above cannot be a set, since assuming it is a set leads to a contradiction. The NBG set theory differentiates between classes and sets. In this set theory, the collection A above is a class which is not a set.

S

satisfiable   Let $\mathcal{L}$ be a first order language, and let Γ be a set of well-formed formulas of $\mathcal{L}$. The set Γ is *satisfiable* if there exists a structure $\mathcal{A}$ for $\mathcal{L}$ and a mapping $s : V \rightarrow A$ such that for each formula $\gamma \in \Gamma$, $\mathcal{A}$ satisfies γ with s. (Here, V is the set of variables of $\mathcal{L}$ and A is the universe of $\mathcal{A}$.)

satisfy   Let $\mathcal{L}$ be a first order language, α be a well-formed formula of $\mathcal{L}$, $\mathcal{A}$ be a structure for $\mathcal{L}$, V be the set of variables of $\mathcal{L}$, and $s : V \rightarrow A$ (i.e., s assigns each variable in the language to some element of the universe of $\mathcal{A}$). The function s can be extended to a function $\overline{s} : T \rightarrow A$ from the set T of all terms of $\mathcal{L}$ into A, by induction, as follows.

(i.) If x is a variable of $\mathcal{L}$, then

$$\overline{s}(x) = s(x) .$$

(ii.) If c is a constant symbol of $\mathcal{L}$, then

$$\overline{s}(c) = c^{\mathcal{A}} ,$$

where $c^{\mathcal{A}}$ is the element of A assigned to c by $\mathcal{A}$.

(iii.) If $t_1, \ldots, t_n$ are terms of $\mathcal{L}$ and f is an n-ary function symbol of $\mathcal{L}$, then

$$\overline{s}(f(t_1, \ldots, t_n)) = f^{\mathcal{A}}(\overline{s}(t_1), \ldots, \overline{s}(t_n)) ,$$

where $f^{\mathcal{A}}$ is the n-ary function on A assigned to f by $\mathcal{A}$.

The structure $\mathcal{A}$ satisfies α with s (notation: $\models_{\mathcal{A}} \alpha[s]$) and is defined by induction on the complexity of α as follows.

(i.) If $\alpha = (t_1 = t_2)$, where t_1 and t_2 are terms of $\mathcal{L}$, then $\mathcal{A}$ satisfies $(t_1 = t_2)$ with s if $\overline{s}(t_1) = \overline{s}(t_2)$.

(ii.) If $\alpha = P(t_1, \ldots, t_n)$, where $t_1, \ldots, t_n$ are terms of $\mathcal{L}$ and P is an n-ary predicate symbol of $\mathcal{L}$, then $\mathcal{A}$ satisfies $P(t_1, \ldots, t_n)$ with s if $(\overline{s}(t_1), \ldots, \overline{s}(t_n)) \in P^{\mathcal{A}}$, where $P^{\mathcal{A}}$ is the n-ary relation on A assigned to P by $\mathcal{A}$.

(iii.) If $\alpha = (\neg \beta)$, then $\mathcal{A}$ satisfies $(\neg \beta)$ with s if $\mathcal{A}$ does not satisfy β with s.

(iv.) If $\alpha = (\beta \rightarrow \gamma)$, then $\mathcal{A}$ satisfies $(\beta \rightarrow \gamma)$ with s if $\mathcal{A}$ satisfies $(\neg \beta)$ with s or $\mathcal{A}$ satisfies γ with s.

(v.) If $\alpha = \forall v \beta$, then $\mathcal{A}$ satisfies $\forall v \beta$ with s if for all $a \in A$, $\mathcal{A}$ satisfies β with the following modified version $s_a : V \rightarrow A$ of s:

$$s_a(x) = \begin{cases} s(x) & \text{if } x \neq v \\ a & \text{if } x = v . \end{cases}$$

Let $a_1, \ldots, a_n \in A$ and let φ be a well-formed formula with free variables from among $v_1, \ldots, v_n$. The notation $\models_{\mathcal{A}} \varphi[a_1, \ldots, a_n]$ means that there is an $s : V \rightarrow A$ with $s(v_i) = a_i$, for $1 \leq i \leq n$, and $\mathcal{A}$ *satisfies* φ with s.

saturated model   A model A that realizes as many types as possible. More precisely, if A is a model in the language L and X is any subset of A, let L_X be the expansion of L which adds a constant symbol c_x for each $x \in X$. Then for a cardinal κ, A is κ-saturated if for any $X \subseteq A$ of size less than κ, every type $\Phi(x)$ in the language L_X which is consistent with the theory of A (using L_X) is realized in A. That is, there is some $a \in A$ such that $A \models \phi(a)$ for every $\phi \in \Phi$.

A model A is saturated if it is $|A|$-saturated. The rationals are a *saturated model* of the theory of dense linear orderings without endpoints.

Schauder Fixed-Point Theorem   Let X be a closed convex subset of a Banach space. Then any continuous map $f : X \rightarrow X$ for which the closure of $f(X)$ is compact must have a fixed point; that is, there is an $x \in X$ with $f(x) = x$. In particular, any continuous mapping from a compact convex subset of a Banach space into itself has a fixed point.

Schröder-Bernstein Theorem   *See* Cantor-Bernstein Theorem.

s-cobordism   A geometric notion of equivalence for piecewise linear manifolds. An h-cobordism is a manifold W with boundary the disjoint union of two manifolds M_0 and M_1, in which the inclusion maps $i_0 : M_0 \longrightarrow W$ and $i_1 : M_1 \longrightarrow W$ are homotopy equivalences. This can be refined using the notion of simple homotopy. If (K,L) is a pair of simplicial complexes with $K = L \cup B$, with B a closed n-cell

1-58488-050-3/01/$0.00+$.50
© 2001 by CRC Press LLC

and $B \cap L$ is a face of B, then K is said to collapse to L, and L expands to K. This generates an equivalence relation on polyhedra called simple homotopy equivalence. An *s-cobordism* is an h-cobordism in which the inclusions i_0 and i_1 are simple homotopy equivalences.

s-Cobordism Theorem   Let W be an s-cobordism, with boundary the disjoint union of two manifolds M_0 and M_1. Then, if the dimension of W is at least 6, W is actually equivalent (as a polyhedral manifold) to the product manifold $M_0 \times [0, 1]$. This would be false if the inclusion maps were only homotopy equivalences.

secondary cohomology operation   An image of a lift of a cohomology class u in $H^i(Y; A)$, formed in the following manner. The class u is represented by a map $u : Y \longrightarrow K(A, i)$. Let α be a cohomology operation corresponding to the map $\alpha : K(A, i) \longrightarrow K(B, j)$ for which $\alpha u = 0$. Let X represent the two-stage Postnikov tower given by α, so that

$$K(B, j - 1) \longrightarrow X \longrightarrow K(A, i) \longrightarrow K(B, j)$$

is a fibration with maps i, p, and α, respectively. Let $\beta : X \longrightarrow K(G, n)$ represent a class in $H^n(X; G)$. Since $\alpha u = 0$, there is a map v such that v composed with the map $X \longrightarrow K(A, i)$ is homotopic to u. The cohomology class in $H^n(Y; G)$ given by composing βu is the *secondary cohomology operation* given by this procedure evaluated on u.

This operation is only determined up to a coset. If everything is done in the stable range, then the indeterminacy is due only to the choice of v; any two choices may differ by any element of $H^n(Y, G)$ which is in the image of $i^*(\alpha) : H^{j-1}(Y; B) \longrightarrow H^n(Y; G)$. One usually only uses secondary operations in the stable range (j and n less than $2i - 1$) because indeterminacy is difficult to determine otherwise.

These are operations that arise from the relations among primary cohomology operations. The Adem relation $Sq^3 Sq^1 + Sq^2 Sq^2 = 0$ generates a secondary cohomology operation that shows that η^2 is essential (not homotopic to zero), where η represents the Hopf map $S^3 \longrightarrow S^2$ (in the Hopf bundle) or any suspension of that map. Note that a secondary cohomology operation is

not defined on the whole cohomology group in general. *See* Adem relations. *See also* primary cohomology operation.

second category space   A topological space X which is not first category; that is, X is not equal to the union of a countable collection of closed subsets with empty interiors.

second countable topological space   A topological space that has a countable basis for its topology. For example, the real line (with its standard topology) is second countable since open intervals with rational endpoints form a basis.

semicircle   An arc of a circle, connecting two points on a diameter, for example $\{(x, y) : x^2 + y^2 = 1, \ y \geq 0\}$.

sentence   A well-formed formula of a first order language having no free variables. *See* free variable.

sentential calculus   *See* propositional calculus.

sentential logic   *See* propositional logic.

separable topological space   A topological space with a countable, dense subset. For example, the real line (with its standard topology) is separable, since the set of rational numbers is countable and dense in the reals.

separated sets   Two subsets A and B of a topological space X which satisfy $\bar{A} \cap B = \bar{B} \cap A = \emptyset$, where $\bar{A}$ and $\bar{B}$ denote the closure of A and B.

separation axioms   A system of axioms for topological spaces X which measure, in increasing fashion, the extent to which points and subsets are separated by the topology on X. The standard axioms are called the T_0, T_1, T_2, T_3, and T_4 axioms. Other axioms, including completely regular, Tychonoff, and Urysohn spaces, refine and extend this list.

separation by a continuous function   The property of a continuous function $f : X \rightarrow$

[0, 1] that, for two subsets $A, B \subset X$, we have $f(A) = \{0\}$ and $f(B) = \{1\}$.

set **(1)** In naive set theory. A *set* is any collection of arbitrary objects. When such a collection is seen as a single entity, it is considered a set. Alternative terms: collection or family (in particular, these terms are often used for sets of sets, sets of sets of sets, and so on; so a set of sets is often called a family of sets, or a collection of sets). Sets are determined by their elements (their members). Standard set notation defines a set by listing or describing its elements within curly brackets: $\{,\}$. For example, the set whose only elements are the number 3 and the letter Q is written in list form as $\{3, Q\}$ or $\{Q, 3\}$ (the order of listing does not matter). The set $\{2, 4, 6, 8\}$ (list form) can also be expressed in description form as $\{2n : n = 1, 2, 3, 4\}$ or as $\{2n \mid n = 1, 2, 3, 4\}$.

(2) In axiomatic set theory. A formal mathematical object whose existence is a consequence of the axiomatic system with which one is working. For example, in Zermelo-Fraenkel set theory, *sets* are built using axioms such as Union, Comprehension, Power Set, etc. *See* Zermelo-Fraenkel set theory, Bernays-Gödel set theory.

set theory **(1)** Axiomatic set theory. The branch of mathematics whose purpose is to study sets within a formal axiomatic framework. Also known as the foundation of mathematics, referring to the notion that all of mathematics can be carried out within *set theory*. For example, Zermelo-Fraenkel set theory models mathematics in a natural way. *See also* Zermelo-Fraenkel set theory, Bernays-Gödel set theory.

(2) Naive set theory. The practice of dealing with sets as arbitrary collections of objects and performing operations on such sets without appealing to axioms.

sexagesimal number system A number system, used by the ancient Babylonian civilization, that was a base 60 positional system, in contrast to the base 10 positional system commonly used today. The value of a particular number depends both on the numerals used in its representation and the placement of these numerals. Using the symbols $|$ and $<$ to represent 1 and 10 respectively, one can denote the number 34

as $<<<$ $||||$, 154 is expressed as $||$ $<<<$ $||||$ $(2 \times 60 + 34)$, and 5000 is represented by $|$ $<<$ $|||$ $<<$ $((1 \times 60^2) + (23 \times 60) + 20)$. A high base such as 60 is useful for dealing with large numbers, since the "place values" represent powers of the base (60), namely, 1, 60, 3600, 216000, ... One of the difficulties is the fact that there must be 60 "digits" (representing the values 0 to 59). In fact, the Babylonians did not have a symbol for zero so the number $<<$ $|||||||$ $<$ $|||$ could represent 1663 $((27 \times 60) + 13)$ or 97213 $(27 \times 3600) + (0 \times 60) + 13)$. *See also* base of number system.

sheaf A structure F on a topological space X, which assigns an object $F(U)$ to each open subset U of X, and for each inclusion U in V of open sets in X, F assigns a restriction map $r_{V,U} : F(V) \longrightarrow F(U)$ so that $r_{U,U}$ is the identity on $F(U)$ and whenever U in V in W are nested open sets, $r_{V,U} \circ r_{W,V} = r_{W,U}$. Further, whenever $U = \cup_{a \in I} U_a$ is a covering of U by open sets U_a, and $\{f_a\}$ is a collection of elements f_a in $F(U_a)$ such that the restrictions of f_a and f_b to $U_a \cap U_b$ are equal, there is a unique element f in $F(U)$ such that the restriction $r_{U,U_a}(f)$ to each U_a is just f_a.

Example: The collection of open sets of a space X is a *sheaf* with $F(U) = U$. One may also use sheaves as coefficients in homology of X.

sieve of Eratosthenes A method (named after the Greek mathematician Eratosthenes) for "sifting" out the prime numbers less than a fixed integer N. It relies on the fact that if n is a positive integer less than or equal to N, then n is either a prime number or has a prime factor that is less than or equal to $\sqrt{N}$.

To find the primes less than or equal to N, first list the integers from 2 to N. Then, circle 2 and cross out all of the other multiples of 2 since we know they cannot be primes (they are divisible by 2). Notice that the smallest integer left that is not circled or crossed out is 3 (the next prime number). Circle 3 and cross out the remaining multiples of 3. Now, circle the smallest integer that is neither circled nor crossed out (5) and cross out all its other multiples. Continue this process until the smallest integer that is neither circled nor crossed out is greater than

$\sqrt{N}$. Circle the remaining integers in the list; the integers that have been circled are the primes less than or equal to N.

simple closed curve   A topological space C that is homeomorphic to the unit circle. Intuitively, this means that C does not cross itself.

simple homotopy equivalence   A homotopy equivalence $f : S_1 \rightarrow S_2$ between two simplicial complexes which is obtained as a composition of elementary contractions and expansions. Given an n-simplex σ of a simplicial complex S such that σ is the face of a unique $n+1$-simplex τ, an elementary contraction of S is the map that collapses σ and τ to a point. An elementary expansion of S is the inverse of an elementary contraction.

simplex   Let $\{a_0, \dots, a_n\}$ be a geometrically independent subset of $\mathbf{R}^n$. The n-simplex σ spanned by $\{a_0, \dots, a_n\}$ is the set of all points

$$x = \sum_{k=0}^{n} t_k a_k, \quad \text{where} \quad \sum_{k=0}^{n} t_k = 1,$$

and $t_k \geq 0$ for all k. The points $\{a_0, \dots, a_n\}$ are called the vertices of σ. The t_k are called the barycentric coordinates for σ. Any simplex spanned by a subset of $\{a_0, \dots, a_n\}$ is called a face of σ. For example, a 0-simplex is a point, a 1-simplex is a line segment, and a 2-simplex is a triangle.

simplicial approximation   Let $f : S_1 \rightarrow S_2$ be a continuous function between simplicial complexes. A simplicial mapping $g : S_1 \rightarrow S_2$ is a *simplicial approximation* for f if $f(\text{St}(v)) \subseteq \text{St}(g(v))$ for every vertex v of S_1 where $\text{St}(v)$ denotes the star of the vertex v.

simplicial complex   A set V of vertices, together with a set K of finite subsets of V called simplices, satisfying the condition: if σ is a simplex and τ is a subset of σ, then τ is also a simplex.

simply connected domain   A subset D of a topological space X which is open, connected, and simply connected as a subspace of X. *See* simply connected space. That is, D must be path-connected and have a trivial fundamental group, $\pi_1(D)$, as a subspace of X.

simply connected space   A topological space X which is path-connected and has trivial fundamental group $\pi_1(X)$. That is, any closed path in X is homotopic to a constant path. For example, a circular disc in the plane is *simply connected,* while an annulus is not because there are paths in it (going around the annulus) which cannot be continuously deformed to a constant path.

singleton set   Any set with exactly one element. For example, $\{7\}$ is a singleton.

singular cardinal   A cardinal number whose cofinality is smaller than itself. Thus, if κ is a *singular cardinal,* κ is not regular and $\text{cf}(\kappa) < \kappa$. For example, $\aleph_\omega$ is a singular cardinal. *Compare with* regular cardinal.

singular complex   For X a topological space, the chain complex $S(X) = \{S_n(X)\}$ of free Abelian groups (or free modules over a ring R), generated by singular simplices. *See* chain complex. The standard k-simplex is the set $\sigma_k = \{(x_0, x_1, \dots, x_k) \in R^{k+1} : x_0 + \dots + x_k = 1,$ each $x_i \geq 0\}$. A singular k-simplex is a continuous function $\phi : \sigma_k \longrightarrow X$. For each $n \geq 0$, $S_n(X)$ is the free module generated by the singular k-simplices. The boundary map $\partial_k : S_k(X) \longrightarrow S_{k-1}(X)$ is constructed by taking a singular simplex ϕ to the alternating sum of the $(k-1) - simplices$ determined by restricting ϕ to its faces.

singular homology   A graded Abelian group $H(X) = \{H_n(X)\}$, determined by a space X. The group $H_k(X)$ is the quotient of the singular cycles $Z_k(X) = \ker \partial_k : S_k(X) \longrightarrow S_{k-1}(X)$ modulo the boundaries $B_k(X) = \partial_{k+1}(S_{k+1}(X))$. The *singular homology* groups are fundamental invariants of X.

singular n-boundary   If $\{S_n(X)\}$ is the singular complex of a space X, then the nth boundary group $B_n(X)$ is the subgroup of $S_n(X)$ consisting of elements of the form ∂c for c in $S_{n+1}(X)$. The elements of $B_n(X)$ are *singular n-boundaries*.

singular n-chain An element of the free Abelian group (or, more generally, the free module over a ring R) $S_n(X)$, a linear combination of singular n-simplices in a topological space X. *See* singular n-simplex.

singular n-simplex The standard n-simplex is the set

$$\sigma_n = \{(x_0, x_1, \ldots, x_n) \in \mathbf{R}^{n+1} :$$

$$x_0 + \ldots + x_n = 1, \text{ each } x_i \geq 0\} .$$

A *singular n-simplex* in a space X is a continuous function $\phi : \sigma_n \longrightarrow X$.

skew lines Two lines that do not meet in projective geometry, which can occur in $\mathbf{P}^n$ for $n \geq 3$ only.

Skolem expansion (1) The *Skolem expansion* of a language L is $L \cup \{f_\phi : \exists x\phi$ is a formula in $L\}$, where each f_ϕ is a Skolem function for ϕ. *See* Skolem function.

(2) The *Skolem expansion* of a theory T in the language L is T together with the set of sentences

$$\forall \bar{y}\big(\exists x\phi(x, \bar{y}) \rightarrow \phi(f_\phi(\bar{y}), \bar{y})\big)$$

for each Skolem function f_ϕ of L. The language of the expanded theory is the Skolem expansion of L.

(3) A *Skolem expansion* of a structure A in the language L is a model A' which adds to A consistent interpretations of the Skolem functions of L. That is, for each Skolem function f_ϕ of L,

$$A' \models \forall \bar{y}\big(\exists x\phi(x, \bar{y}) \rightarrow \phi(f_\phi(\bar{y}), \bar{y})\big) .$$

The language of the expanded model A' is the Skolem expansion of L.

Skolem function If $\exists x\phi(x, \bar{y})$ is a formula with all its free variables in $\bar{y} = \{y_1, \ldots, y_n\}$, then a *Skolem function* for ϕ, f_ϕ, satisfies

$$\forall \bar{y}\big(\exists x\phi(x, \bar{y}) \rightarrow \phi(f_\phi(\bar{y}), \bar{y})\big) .$$

In effect, the symbol $f_\phi(\bar{y})$ names a witness of the existential statement $\exists x\phi(x, \bar{y})$ for each $\bar{y}$ which has one.

Skolem hull If X is a subset of an L-structure A, the *Skolem hull* of X is the smallest submodel of the Skolem expansion of A which contains X. Equivalently, it is the smallest subset of A containing X which is closed under the operations of the Skolem expansion. Any nonempty Skolem hull is an elementary submodel of the original model A. *See* Skolem expansion.

Skolem normal form A formula ψ is in *Skolem normal form* if $\psi = \forall \bar{x}\exists \bar{y}\phi(\bar{x}, \bar{y})$, where ϕ is quantifier-free.

Skolem theory A theory T in the language L which is its own Skolem expansion; that is, T contains

$$\forall \bar{y}\big(\exists x\phi(x, \bar{y}) \rightarrow \phi(f_\phi(\bar{y}), \bar{y})\big)$$

for each Skolem function f_ϕ of L. *See* Skolem function.

smoothing A smooth structure on a topological manifold M, which induces the given topological structure. A *smoothing* of a piecewise linear manifold is a smooth structure in which each simplex is smooth.

smoothing problem The problem of determining the existence or non-existence of a smoothing of a topological or piecewise linear manifold M. *See* smoothing. The problem always has an affirmative solution in dimensions less than or equal to three, but there are many counterexamples in higher dimensions, both for existence and uniqueness of smoothings.

smooth manifold A real manifold whose transition functions are smooth, or $C^{(k)}$-differentiable, for $k \geq 1$. Namely, a space M with an open covering $\{U_\alpha\}$ and identifications $\phi_\alpha : U_\alpha \rightarrow \mathbf{R}^n$, where n is the dimension of the manifold and the transition functions $\phi_{\alpha\beta} : U_\alpha \cap U_\beta \rightarrow U_\alpha \cap U_\beta$ are such that $\phi_\alpha = \phi_\beta \circ \phi_{\alpha\beta}$ where they are all defined.

smooth structure A maximal collection of local coordinate systems on a manifold with the property that the coordinate transformation between any two overlapping coordinate systems is differentiable with differentiable inverse.

Sorgenfrey line   The real line with the topology given by taking the collection of all half-open intervals $[a, b)$ (or $(a, b]$) as a basis. It is also known as the lower (or upper) limit topology.

The *Sorgenfrey line* is normal and Lindelöf but not second countable. Its product with itself (the Sorgenfrey plane) is neither normal nor Lindelöf. Thus, it is an example showing that the product of normal spaces need not be normal, and the product of Lindelöf spaces need not be Lindelöf. *See* normal space.

Sorgenfrey plane   *See* Sorgenfrey line.

space of complex numbers   The complex numbers, visualized as a plane with real and imaginary axes, together with the usual (product) topology of the plane, is a topological space. The set of purely imaginary numbers forms a subspace homeomorphic to the real line.

The imaginary axis, considered a subspace, is homeomorphic to the real numbers.

space of imaginary numbers   *See* space of complex numbers.

space of irrational numbers   A subspace of the space of real numbers: closeness, as described by open sets, is determined by open intervals in the real numbers intersected with the respective set. The space is dense in the space of real numbers; that is, its closure is the space of real numbers.

space of rational numbers   A subspace, usually denoted $\mathbf{Q}$ or $\mathbb{Q}$, of the space of real numbers: closeness, as described by open sets, is determined by open intervals in the real numbers intersected with the respective set. The space $\mathbf{Q}$ is dense in the space of real numbers; that is, its closure is the space of real numbers.

space of real numbers   The set of real numbers together with the usual real line topology generated by open intervals, usually denoted $\mathbf{R}$, $\mathbb{R}$, $\mathbf{R}^1$ or $\mathbf{E}^1$. Intuitively, open sets describe closeness, and typical uses of the real numbers require a topology where decreasing intervals around a point describe points strictly closer to that point. $\mathbf{R}$ is also a metric space with distance function $d(x, y) = |x - y|$. *See also* real number.

span   The smallest subspace of a vector space F containing a given set C of vectors in F.

sphere   **(1)** The subspace S^n of $\mathbf{R}^{n+1}$ consisting of all points $(x_1, \ldots, x_{n+1})$ with $x_1^2 + \cdots + x_{n+1}^2 = 1$.

(2) More generally, a space homeomorphic to S^n.

spherical distance   The greatest lower bound of the lengths of all paths between two points p and q lying on the (unit) sphere. It is the length of the short great circle arc joining p to q.

spherical polygon   A closed curve on the surface of the sphere made up of a finite number of great circle arcs.

spherical triangle   A closed curve consisting of three points A, B, and C on the sphere, together with a great circle arc joining each pair of points. Sometimes the arcs are required to be shortest arcs.

square-free integer   An integer that is not divisible by any perfect square other than 1. The prime factorization of a square-free integer contains no exponent greater than 1. Thus, 21 is square-free, but 20 is not, since 2^2 is a divisor of 20.

square number   An integer that equals n^2 for some integer n.

square root   **(1)** (Of a non-negative real number r) The unique non-negative real number s so that $s^2 = r$, denoted $\sqrt{r}$.

(2) If z and w are complex numbers such that $w^2 = z$, then w is said to be a *square root* of z (there will be two square roots of a given *nonzero* complex number, since if w is a square root of z, so is $-w$ and by the Fundamental Theorem of Algebra, the equation $x^2 = z$ has at most two distinct solutions).

stable (primary) cohomology operation   Let ΣX denote the suspension of a space X ($S^1 \wedge X$). Then $H^q(X)$ is isomorphic to $H^{q+1}(\Sigma X)$ by an isomorphism called the suspension isomorphism (here denoted Σ), natural in X.

A cohomology operation P is stable when $\Sigma P = P\Sigma$, that is, P commutes with the suspension isomorphism. The Steenrod square and power operations are examples of *stable (primary) cohomology operations*. *See* Steenrod square operation.

stable range   Some algebraic invariants behave well with respect to suspension, sometimes with connectivity restrictions. For example, if X is $(n-1)$-connected and $i \leq 2n - 2$, then $\pi_i(X)$ is isomorphic to $\pi_{i+1}(\Sigma X)$. This range is called the *stable range* of X.

In homotopy theory, one may be concerned with the stable range in calculating homotopy groups or the effect of cohomology operations.

stable secondary cohomology operation   *See* stable (primary) cohomology operation.

stably parallelizable manifold   A smooth manifold M such that the Whitney sum of the tangent bundle of M and a trivial bundle over M is a trivial bundle. For example, the tangent bundle of the sphere S^2 is not trivial, but its Whitney sum with a 1-dimensional trivial bundle is trivial. Thus, S^2 is stably parallelizable.

stationary set   A set of ordinals $S \subseteq \kappa$ which meets every closed unbounded set in κ; i.e., $S \cap C \neq \emptyset$ for each closed and unbounded $C \subseteq \kappa$. *Stationary sets* are somewhat large; for example, they are unbounded because for each $\alpha < \kappa$, $C_\alpha = [\alpha, \kappa)$ is closed and unbounded.

Steenrod algebra   The algebra of all cohomology operations for ordinary mod p cohomology, for a prime p. When $p = 2$, the Steenrod square operations Sq^i generate the *Steenrod algebra*. For odd primes p, the analog of the squares are the pth power operations P^i; these together with the Bockstein operation generate the Steenrod algebra for p odd.

The Steenrod squares are defined as additive cohomology operations

$$Sq^i : H^q(X, A) \longrightarrow H^{q+i}(X, A)$$

(additive natural transformations

$$Sq^i : H^q(-) \longrightarrow H^{q+i}(-))$$

for which

(i.) Sq^0 is the identity;
(ii.) for u in $H^i(X, A)$, $Sq^i u = u \cup u = u^2$;
(iii.) for u in $H^i(X, A)$ and $i > k$, $Sq^k u = 0$;
(iv.) for u in $H^a(X, A)$ and v in $H^b(Y, B)$, the effect on uv in $H^*(X \times Y, A \times B)$ is given by the Cartan formula:

$$Sq^k(u \times v) = \sum_{i+j=k} Sq^i u \times Sq^j v .$$

The power operations are defined by similar properties. *See also* Bockstein operation.

The cohomology of a space is a comodule over the Steenrod algebra. This structure (coaction) is preserved by many (not all) constructions and calculational techniques, and hence can be used to calculate the cohomology of certain spaces, for example, Eilenberg MacLane spaces (whose cohomology can be calculated using the Serre spectral sequence).

Steenrod pth power operation   *See* Steenrod algebra.

Steenrod square operation   *See* Steenrod algebra.

stereographic projection   An identification of the plane with the sphere minus a point, N, say, obtained by projecting from N a point P on the sphere different from N. If the sphere of radius 1 touches the (x, y) coordinate plane at the origin and $N = (0, 0, 2)$, the projection sends (x, y, z) to $(\frac{2x}{2-z}, \frac{2y}{2-z})$.

Sometimes, instead, the entire sphere is identified with the complex plane, together with the point at infinity. *See also* complex sphere.

Stirling number   The number S_n^m for $n \geq m$, which denotes the number of partitions of a set of n objects into m non-empty subsets. These numbers are given by the recurrence relations $S_n^1 = 1 = S_n^n$ and $S_{n+1}^k = S_n^{k-1} + k S_n^k$ for $1 < k < n$.

Stone-Čech compactification   The unique largest compactification $\beta(X)$ of a completely regular topological space, X. Its usefulness derives from the fact that any continuous function from X to a compact Hausdorff space may be extended uniquely and continuously to $\beta(X)$.

To construct $\beta(X)$, let $\mathcal{F}$ be the set of all continuous functions from X to the closed unit interval, $[0, 1]$. Then the product space $[0, 1]^{\mathcal{F}}$, of one copy of the unit interval for each $f \in \mathcal{F}$, is a compact Hausdorff space by Tychonoff's Theorem. Imbed X in $[0, 1]^{\mathcal{F}}$ by mapping $x \in X$ to the element of the product with $f(x)$ in its f-coordinate. $\beta(X)$ is the closure of the image of X under this imbedding. *See also* one-point compactification.

strong induction   A method of proof over well-ordered sets. In practice, *strong induction* is typically used over the set of natural numbers. Strong induction has a base-case, like induction, but a different inductive step. Expressed in formal notation, the base-case is $P(n_0)$, for some n_0; the inductive step has the form:

$$(\forall k)[[(\forall n \leq k)P(n)] \rightarrow P(k+1)] .$$

From these the conclusion is $(\forall k \geq n_0)P(k)$, where $P(k)$ is some statement and n_0, n, k are natural numbers. Strong induction is equivalent to induction. *See* induction.

strongly multiplicative function   A multiplicative function f having the property that $f(p^i) = f(p)$ for all primes p and all positive integers i. For example, the function $f(n) = \frac{\phi(n)}{n}$, where ϕ is the Euler phi function, is strongly multiplicative. *See* multiplicative function, Euler phi function. *See also* completely multiplicative function.

strong pseudoprime   *See* pseudoprime.

structure   A mapping $\mathcal{A}$, which assigns values to the quantifier symbol, the predicate symbols, the constant symbols, and the function symbols of a first order language $\mathcal{L}$, as follows.

(i.) $\mathcal{A}$ assigns to the quantifier symbol $\forall$ a nonempty set A (sometimes denoted by $|\mathcal{A}|$), called the universe of $\mathcal{A}$.

(ii.) For each n-ary predicate symbol P, $\mathcal{A}$ assigns P to an n-ary relation $P^{\mathcal{A}} \subseteq A^n$.

(iii.) For each constant symbol c, $\mathcal{A}$ assigns c to an element $c^{\mathcal{A}}$ of A.

(iv.) For each n-ary function symbol f, $\mathcal{A}$ assigns f to an n-ary function $f^{\mathcal{A}} : A^n \rightarrow A$.

For example, if $\mathcal{L}$ is the language of elementary number theory (*see* first order language),

then one possible structure for $\mathcal{L}$ is the intended structure $\mathcal{N}$, which assigns the quantifier $\forall$ to $\mathbf{N}$, the set of natural numbers, and $<, 0, S, +, \cdot, E$ to their intended interpretations on $\mathbf{N}$. There are other (non-standard) structures for this language.

A *structure* is sometimes called a model.

subbasis for a topology   A collection of subsets of a topological space X whose set of finite intersections forms a basis for the topology τ of X. For example, the set of all open intervals of the form $(-\infty, a)$ or (a, ∞) is a subbasis for the usual topology on $\mathbf{R}$ because each basic open set (a, b) can be written as $(-\infty, b) \cap (a, \infty)$.

Any collection S of subsets of a nonempty set X generates a topology on X by declaring S to be a subbasis. That is, the topology is the set of all unions of finite intersections of elements from S. The topology generated in this way is the smallest topology on X which contains S.

subbundle   A bundle $F' \longrightarrow E' \longrightarrow B$ contained a given bundle $F \longrightarrow E \longrightarrow B$.

The tangent bundle and normal bundle of a manifold M embedded in $\mathbf{R}^n$ are both *subbundles* of the trivial bundle $M \times \mathbf{R}^n$.

subcategory $\mathcal{C}'$ is a *subcategory* of a category $\mathcal{C}$ if (i.) every object of $\mathcal{C}'$ is an object of $\mathcal{C}$, (ii.) for every pair of objects A, B of $\mathcal{C}'$, if $f : A \rightarrow B$ is a morphism of $\mathcal{C}'$, then f is a morphism of $\mathcal{C}$, and (iii.) for every pair f, g of morphisms of $\mathcal{C}'$, the compositions $f \circ_{\mathcal{C}'} g$ and $f \circ_{\mathcal{C}} g$ are the same morphisms in $\mathcal{C}$ and $\mathcal{C}'$. $\mathcal{C}'$ is a *full subcategory* of $\mathcal{C}$ if, in addition, for every pair A, B of objects of $\mathcal{C}'$, $f : A \rightarrow B$ is a morphism of $\mathcal{C}'$ if and only if f is a morphism of $\mathcal{C}$. For example, the category of sets and bijective functions is a subcategory of the category of sets and injective functions; the category of Abelian groups and group homomorphisms is a full subcategory of the category of groups and group homomorphisms.

subobject   If A is an object of a category $\mathcal{C}$, a *subobject* of A is an ordered pair (f, A'), where A' is an object of $\mathcal{C}$ and f is a monomorphism $f : A' \rightarrow A$. For example, in the category of groups and group homomorphisms, a subobject of the additive group $\mathbf{Z}$ is $(f, \mathbf{E})$, where $\mathbf{E}$ is the additive group of even integers and f is the inclusion map $f : \mathbf{E} \to \mathbf{Z}$. The dual notion of subobject is the notion of quotient object. *See* quotient object.

subset (of a set)   A set S is a *subset* of a set X if all elements of S are also elements of X. If S is a subset of X, the notation is $S \subseteq X$, or sometimes $S \subset X$. For example, $\{4, -2\} \subseteq \{-2, 5, 4\}$. Every set is a subset of itself. *Compare with* proper subset.

subspace   Any subset of a topological space X, with the relative topology inherited from X. *See* relative topology. For example, besides containing all its open subintervals, the subspace topology on the unit interval $[0, 1]$ also includes the half-open intervals $[0, b)$ with $b \leq 1$ and $(a, 1]$ with $a \geq 0$.

substructure   The structure $\mathcal{A}$ for the first order language $\mathcal{L}$ is a *substructure* of the structure $\mathcal{B}$ for $\mathcal{L}$ (notation: $\mathcal{A} \subseteq \mathcal{B}$) if

  (i.) $A \subseteq B$, where A and B are the universes of $\mathcal{A}$ and $\mathcal{B}$, respectively,

  (ii.) for each n-ary predicate symbol P, the n-ary relation $P^{\mathcal{A}}$ is the restriction of $P^{\mathcal{B}}$ to A^n; i.e., $P^{\mathcal{A}} = P^{\mathcal{B}} \cap A^n$,

  (iii.) for each constant symbol c, $c^{\mathcal{A}} = c^{\mathcal{B}}$, and

  (iv.) for each n-ary function symbol f, $f^{\mathcal{A}}$ is the restriction of $f^{\mathcal{B}}$ to A^n.

If $\mathcal{A}$ is a substructure of $\mathcal{B}$, then $\mathcal{B}$ is called an *extension* of $\mathcal{A}$.

Sometimes the term submodel is synonymous with substructure.

successor cardinal   A cardinal number κ such that there exists some other cardinal λ such that $\lambda^+ = \kappa$. For example, $\aleph_{17}$ is a *successor cardinal* since $\aleph_{16}^+ = \aleph_{17}$; $\aleph_{17}$ is a limit ordinal. *Compare with* limit cardinal, successor ordinal.

successor of a cardinal   If κ is a cardinal, the *cardinal successor* of κ, denoted by κ^+, is the least cardinal that is greater than κ. For example, $3^+ = 4$ and $\aleph_0^+ = \aleph_1$. *Compare with* successor of an ordinal.

successor of an ordinal   If α is an ordinal, the *ordinal successor* of α, denoted by $\alpha + 1$, is $\alpha \cup \{\alpha\}$; it is the least ordinal that is greater than α. For example, $3 + 1 = 4$ and the ordinal successor of ω is $\omega + 1$. *Compare with* successor of a cardinal.

successor of a set   If S is any set, its *successor* is $S \cup \{S\}$.

successor ordinal   An ordinal number α such that there exists some other ordinal β such that $\beta + 1 = \alpha$. For example, $\omega^3 + 5$ is a *successor ordinal* since $(\omega^3 + 4) + 1 = \omega^3 + 5$. The cardinal number $\aleph_1$ is a successor cardinal but not a successor ordinal. *Compare with* successor cardinal, limit ordinal.

sum of cardinal numbers   The cardinal number that is equinumerous with the disjoint union of the summands. For example, $1 + 1 = 2$, and $\aleph_0 + \aleph_3 = \aleph_3$.

sum of divisors function   The arithmetic function, denoted σ, which, for any positive integer n, returns the sum of the positive divisors of n, i.e., $\sigma(n) = \sum_{d|n} d$. (*See* arithmetic function.) For example, $\sigma(10) = 1 + 2 + 5 + 10 = 18$. It is multiplicative; its value at a prime power is given by

$$\sigma(p^i) = \frac{p^{i+1} - 1}{p - 1} .$$

See also multiplicative function, sum of kth powers of divisors function.

sum of kth powers of divisors function
The family of arithmetic functions, denoted σ_k, which, for any positive integer n and a fixed nonnegative integer k, returns the sum of the kth powers of the positive divisors of n, i.e., $\sigma_k(n) = \sum_{d|n} d^k$. (*See* arithmetic function.) For example,

$$\sigma_2(8) = 1^2 + 2^2 + 4^2 + 8^2 = 85 .$$

The function σ_0 is the number of divisors function τ, and σ_1 is the sum of divisors function σ. The functions σ_k are all multiplicative; their value at a prime power is given by

$$\sigma_k(p^i) = \frac{p^{k(i+1)} - 1}{p^k - 1} .$$

surd   Another name for the radical sign $\sqrt{}$. *See also* radical.

surjection   A function $f : A \to B$ such that the image (range) of f is all of B; that is, for any $b \in B$ there is an $a \in A$ with $f(a) = b$.

Suslin line   A dense linear ordering $(L, <)$ which in the order topology has the countable chain condition but is not separable. That is, L has no countable dense subset, but any collection of pairwise disjoint nonempty open sets in L is countable.

It is possible to characterize the real line **R** as the unique dense linear order without endpoints which is complete and separable. The question arose as to whether separability could be replaced by the countable chain condition, and so the existence of a *Suslin line* would mean that this new set of conditions does not characterize **R**. However, the existence of a Suslin line is independent of the axioms of set theory, and thus, so is the characterization. *See also* Suslin tree.

Suslin's hypothesis   The assertion that there is no Suslin line. *See* Suslin line. That is, there is no dense linear ordering which in the order topology has the countable chain condition but is not separable. *Suslin's hypothesis* (abbreviated SH) is independent of the axioms of set theory: it is a consequence of Martin's Axiom, but ¬SH is a consequence of Diamond ($\diamondsuit$), which holds in the constructible universe. These results are usually obtained indirectly, by considering Suslin trees rather than Suslin lines. *See also* Martin's Axiom.

Suslin tree   A tree of height ω_1, which has no uncountable antichains or branches. That is, any subset $A \subseteq T$ consisting of incomparable elements (antichain) or any set $B \subseteq T$ totally ordered by $<$ (branch) must be countable. The existence of *Suslin trees* is independent of the axioms of set theory. In fact, Suslin trees provide a way to prove the independence of Suslin's hypothesis (SH) because a Suslin tree exists if and only if a Suslin line exists. *See* Suslin line.

For any infinite cardinal κ, a κ-Suslin tree is a tree of height κ in which all antichains and branches have size less than κ.

See also Aronszajn tree, Kurepa tree.

symmetric difference   The *symmetric difference* of two sets A and B, written $A \triangle B$, is the set $(A \setminus B) \cup (B \setminus A)$. That is, it is the set of all elements that belong to either A or B but not both.

symmetric relation   A binary relation R such that $(x, y) \in R$ implies $(y, x) \in R$, for all x, y. For example, the equality relation is symmetric.

T

T$_0$ space   A topological space X such that, for any two distinct points of X, there is a neighborhood of one which does not contain the other. That is, for all x, y in X, with $x \neq y$, there is an open set U such that either $x \in U$ and $y \notin U$, or $y \in U$ and $x \notin U$. T_0 *spaces* are also known as Kolmogorov spaces.

T$_1$ space   A topological space X such that, for any two distinct points of X, there are neighborhoods of both which do not contain the other. That is, for all x, y in X, with $x \neq y$, there are open sets U and V such that $x \in U$ and $y \notin U$, while $y \in V$ and $x \notin V$. This is equivalent to each singleton $\{x\}$ being closed in X.

T$_2$ space   A topological space X such that any two distinct points can be separated by open sets. That is, for all x, y in X, with $x \neq y$, there are open sets U and V such that $x \in U$, $y \in V$, and $U \cap V = \emptyset$. *See also* Hausdorff topological space.

T$_{3\frac{1}{2}}$ space   A topological space X such that X is a T_1 space and points and closed sets in X can be separated by continuous functions. That is, for all closed $C \subseteq X$ and $x \notin C$, there is a continuous $f : X \rightarrow [0, 1]$ such that $f(x) = 0$ and $f(c) = 1$ for all $c \in C$. Including the condition T_1 ensures $T_{3\frac{1}{2}} \subseteq T_3$. *See also* completely regular topological space, T_1 space.

T$_3$ space   A topological space X which is a T_1 space and such that points and closed sets can be separated by open sets. That is, for all closed $C \subseteq X$ and $x \notin C$, there are open sets U and V such that $x \in U$, $C \subseteq V$, and $U \cap V = \emptyset$. Including the condition T_1 ensures $T_3 \subseteq T_2$. *See also* regular topological space, T_1 space.

T$_4$ space   A topological space X which is a T_1 space and such that disjoint closed sets can be separated by open sets. That is, for all closed C and D contained in X, if $C \cap D = \emptyset$, then there

are open sets U and V such that $C \subseteq U$, $D \subseteq V$, and $U \cap V = \emptyset$. Including the condition T_1 ensures $T_4 \subseteq T_{3\frac{1}{2}}$. *See also* normal topological space, T_1 space.

T$_5$ space   A topological space X which is a T_1 space and such that any two separated sets can be separated by disjoint open sets. That is, for all subsets A and B of X, if

$$A \cap \overline{B} = \overline{A} \cap B = \emptyset$$

(A and B are separated), then there are open sets U and V with $A \subseteq U$, $B \subseteq V$, and $U \cap V = \emptyset$. Including the condition T_1 ensures $T_5 \subseteq T_4$. For any T_1 space X, T_5 is equivalent to hereditary normality. *See also* T_1 space.

tautology   In propositional (sentential) logic, a well-formed propositional formula is a *tautology* if it is true under every truth assignment for the sentence symbols in the formula. For example, if A and B are sentence symbols, then

$$\neg(A \vee B) \leftrightarrow ((\neg A) \wedge (\neg B)),$$

(which is one of DeMorgan's Laws) is a tautology.

In first order logic, let $\mathcal{L}$ be a first order language. A tautology is any well-formed formula of $\mathcal{L}$ which is obtained from a tautology of propositional logic by replacing each sentence symbol in the tautology with a well-formed formula of $\mathcal{L}$.

term   Let $\mathcal{L}$ be a first order language. The set of *terms* of $\mathcal{L}$ is defined inductively as follows.

(i.) If c is a constant symbol of $\mathcal{L}$, then c is a term.

(ii.) If v is a variable of $\mathcal{L}$, then v is a term.

(iii.) If f is an n-place function symbol of $\mathcal{L}$ and $t_1, \ldots, t_n$ are terms of $\mathcal{L}$, then $f(t_1, \ldots, t_n)$ is a term of $\mathcal{L}$.

(iv.) The set of terms is generated by rules (i.), (ii.), and (iii.).

For example, in the first order language of elementary number theory (*see* first order language), $S(0)$ is a term (which is intended to name the natural number 1).

terminal object   An object A of a category $\mathcal{C}$ such that, for every object B of $\mathcal{C}$, there is exactly one morphism f of C such that $f : B \to A$. For example, in the category of sets and functions, a singleton is a *terminal object*. The dual notion of terminal object is *initial object*.

terminating decimal   A decimal representation

$$\ldots a_4 a_3 a_2 a_1 a_0 . a_{-1} a_{-2} a_{-3} \ldots$$

of a real number such that there is an integer N with $a_{-n} = 0$ for all $n \geq N$. A real number r has a terminating decimal representation if and only if there is an integer a and a nonnegative integer N so that $r = \frac{a}{10^N}$. Clearly, any real number with a terminating decimal representation is therefore a rational number.

ternary number system   The real numbers in base $b = 3$ notation. *See* base of number system.

theorem   In first order logic, let $\mathcal{L}$ be a first-order language, and consider a particular predicate calculus for $\mathcal{L}$. Let α be a well-formed formula of $\mathcal{L}$. Then α is a *theorem* of (or is deducible from) the predicate calculus (notation: $\vdash \alpha$) if there is a proof of α in the predicate calculus. *See* proof. If Γ is a set of well-formed formulas of $\mathcal{L}$, then α is a theorem of (or is deducible from) Γ (in the predicate calculus) if there is a proof of α from Γ (notation: $\Gamma \vdash \alpha$).

The notion of theorem in propositional logic is entirely analogous.

theory   A set T of sentences of a first order language $\mathcal{L}$ which is closed under logical implication; i.e., if σ is a sentence of $\mathcal{L}$ which is a logical consequence of T, then $\sigma \in T$ (in notation, $T \models \sigma$ implies $\sigma \in T$). Equivalently, T is a theory if it is closed under deduction; i.e., if σ is provable from T, then $\sigma \in T$ (in notation, $T \vdash \sigma$ implies $\sigma \in T$).

For some authors, the word *theory* simply means a set of sentences, and the notion above is that of a *closed* theory.

Let $\mathcal{A}$ be a structure for $\mathcal{L}$. The theory of $\mathcal{A}$ is the set of sentences of $\mathcal{L}$ which are true in $\mathcal{A}$ (i.e., the theory of $\mathcal{A}$ is the set of sentences σ such that $\mathcal{A}$ is a model of σ). The theory of $\mathcal{A}$ is denoted $Th(\mathcal{A})$ and is a complete theory. *See* complete theory.

Thom complex   Let $E \longrightarrow M$ be a real vector bundle on a manifold M. There is a disk bundle $D \longrightarrow M$ which is given by the open unit disk in each fiber of the vector bundle E. The Thom construction is formed from $E \longrightarrow M$ by identifying all points in E outside of D to a single point, called the point at infinity.

Example: Consider the Möbius band as a line bundle over the circle S^1. The *Thom complex* of this bundle is the real projective plane.

This construction is used in the calculation of cobordism groups. See R. Stong, *Notes on Cobordism Theory*, Princeton University Press, Princeton, NJ, 1968.

topological dimension   Let X be a topological space. The *topological dimension* of X is the smallest non-negative integer n such that, for every open cover $\mathcal{A}$ of X, there is an open cover $\mathcal{B}$ that refines $\mathcal{A}$ (i.e., $\mathcal{A} \subseteq \mathcal{B}$), with the property that some point of X lies in an element of $\mathcal{B}$ and no point of X lies in more than $n + 1$ elements of $\mathcal{B}$.

topological group   A topological space which is also a group such that the inverse and product maps are both continuous. That is, the maps $g \to g^{-1}$ from G to G and $(g_1, g_2) \to g_1 g_2$ from $G \times G$ to G are continuous.

Any discrete group is considered to be a *topological group* with the discrete topology that states: any single element subset is an open set.

topological invariant   A property preserved by homeomorphisms. That is, P is a *topological invariant* if, given any homeomorphism $f : X \to Y$, the space X has property P if and only if Y has property P. For example, connectedness, separability, and normality are all topological invariants.

total ordering   *See* linear ordering.

totient function   *See* Euler phi function.

transcendental number   *See* algebraic number.

transfinite cardinal   Any infinite cardinal number. For example, $\aleph_3$ is a *transfinite cardinal*.

transfinite induction   A method of proof. Suppose $P(\alpha)$ is some statement that describes a property of α, where α is an ordinal. Suppose that all of the following conditions hold: (i.) $P(\alpha_0)$, for some α_0, (ii.) $P(\alpha)$ implies $P(\alpha+1)$, for all $\alpha \geq \alpha_0$, and (iii.) $(\forall \beta < \lambda)\, P(\beta)$ implies $P(\lambda)$, for any nonzero limit ordinal λ. From these three, the conclusion is that $P(\alpha)$ holds for all ordinals $\alpha \geq \alpha_0$. *Transfinite induction* is a generalization of induction.

transfinite ordinal   Any ordinal that is infinite. For example, $\omega+3$ is a *transfinite ordinal*.

transfinite recursion   A method of defining some function; also known as *definition by transfinite recursion*, or sometimes as *definition by transfinite induction*. For any function g on the universe of sets, there exists a unique function f on the class of ordinals such that $f(\alpha) = g(f|\alpha)$, for all ordinals α. *See also* recursion.

transitive relation   A binary relation R such that $[(x, y) \in R] \wedge [(y, z) \in R]$ implies $(x, z) \in R$, for all x, y, z. For example, $\leq$ is a transitive relation on $\mathbf{N}$ since if $n \leq m$ and $m \leq k$, then $n \leq k$.

transitive set   A set A such that, whenever $B \in A$, then $B \subseteq A$.

tree   A partial order $(T, \leq)$ in which, for any $t \in T$, the set of predecessors of t, $\{s \in T : s < t\}$, is well ordered by $<$. That is, any nonempty subset of $\{s \in T : s < t\}$ has a least element. An example of a *tree* is the set of all finite sequences of natural numbers, ordered by extension: $s < t$ if t extends s. Other examples include Aronszajn trees, Kurepa trees, and Suslin trees. *See* Aronszajn tree, Kurepa tree, Suslin tree.

triangular number   The integers in the sequence $1, 3, 6, 10, \ldots$ (which represent the number of lattice points in the plane that lie on the perimeter of isosceles right triangles having integer length legs).

The *triangular numbers* are integers of the form $\sum_{k=1}^{n} k$.

truth assignment   In propositional logic, a function $v : S \to \{T, F\}$ mapping a set S of sentence symbols to $\{T, F\}$, where T is interpreted as true and F is interpreted as false. For example, if $S = \{A_1, A_2, A_3\}$, then a possible truth assignment would be $v : S \to \{T, F\}$ by $v(A_1) = F$, $v(A_2) = T$, and $v(A_3) = T$. Note that there are eight possible *truth assignments* for this particular set of sentence symbols, since there are two choices (T or F) for each value of the function on an element of S. In general, if S has n sentence symbols, then there are 2^n possible truth assignments on S.

A truth assignment $v : S \to \{T, F\}$ is extended using a recursive definition to a truth assignment $\bar{v}$ on the set $\bar{S}$ of all well-formed propositional formulas α which have sentence symbols from S, as follows:
(i.) If α is a sentence symbol in S, then
$$\bar{v}(\alpha) = v(\alpha).$$
(ii.) If $\alpha = (\neg\beta)$, then
$$\bar{v}(\alpha) = \begin{cases} T & \text{if } \bar{v}(\beta) = F \\ F & \text{if } \bar{v}(\beta) = T. \end{cases}$$
(iii.) If $\alpha = (\beta \wedge \gamma)$, then
$$\bar{v}(\alpha) = \begin{cases} T & \text{if } \bar{v}(\beta) = \bar{v}(\gamma) = T \\ F & \text{otherwise.} \end{cases}$$
(iv.) If $\alpha = (\beta \vee \gamma)$, then
$$\bar{v}(\alpha) = \begin{cases} T & \text{if } \bar{v}(\beta) = T \text{ or } \bar{v}(\gamma) = T \\ F & \text{otherwise.} \end{cases}$$
(v.) If $\alpha = (\beta \to \gamma)$, then
$$\bar{v}(\alpha) = \begin{cases} T & \text{if } \bar{v}(\beta) = F \text{ or } \bar{v}(\gamma) = T \\ F & \text{otherwise.} \end{cases}$$
(vi.) If $\alpha = (\beta \leftrightarrow \gamma)$, then
$$\bar{v}(\alpha) = \begin{cases} T & \text{if } \bar{v}(\beta) = \bar{v}(\gamma) \\ F & \text{if } \bar{v}(\beta) \neq \bar{v}(\gamma). \end{cases}$$

truth table   A table of truth values for a well-formed propositional formula α, based on assignments of truth values for the sentence symbols in α. In general, if there are n sentence symbols in α, then the *truth table* will have 2^n rows. The truth tables for the formulas built up from the logical connectives (here A and B are well-formed propositional formulas) are as follows, where T is interpreted as true and F is interpreted as false.

A	$(\neg A)$
T	F
F	T

A	B	$(A \wedge B)$
T	T	T
T	F	F
F	T	F
F	F	F

A	B	$(A \vee B)$
T	T	T
T	F	T
F	T	T
F	F	F

A	B	$(A \to B)$
T	T	T
T	F	F
F	T	T
F	F	T

A	B	$(A \leftrightarrow B)$
T	T	T
T	F	F
F	T	F
F	F	T

A truth table for the more complicated well-formed propositional formula $((A \vee B) \to C)$, where A, B, C are sentence symbols, is as follows.

A	B	C	$(A \vee B)$	$((A \vee B) \to C)$
T	T	T	T	T
T	T	F	T	F
T	F	T	T	T
T	F	F	T	F
F	T	T	T	T
F	T	F	T	F
F	F	T	F	T
F	F	F	F	T

tubular neighborhood   A *tubular neighborhood* of a simple closed curve $L \subset S^3$ is a neighborhood of L homeomorphic to $L \times B^2$ where $L \times \{0\}$ is identified with the curve L.

More generally, a tubular neighborhood of an l-dimensional submanifold $L \subset M$ in an n-dimensional manifold M is a neighborhood of L homeomorphic to $L \times B^{m-l}$.

Turing complete set   A set A of natural numbers which is recursively enumerable and, for any recursively enumerable set B, $B \leq_T A$; i.e., B is computable, relative to A.

An example of a *Turing complete set* is the halting set $K = \{e : \varphi_e(e) \text{ is defined}\}$, where φ_e denotes the partial recursive function with Gödel number e.

Turing complete is sometimes simply referred to as complete.

Turing equivalent   Two sets A and B of natural numbers such that A is Turing reducible to B ($A \leq_T B$) and B is Turing reducible to A ($B \leq_T A$). Intuitively, Turing equivalent sets are sets that code the same information. Turing equivalence (notation: $A \equiv_T B$) is an equivalence relation on the class of all sets of natural numbers. The equivalence classes of $\equiv_T$ are called Turing degrees, or degrees of unsolvability.

As an example, any two Turing complete sets are *Turing equivalent.*

Turing reducibility   Let φ be a partial function on $\mathbf{N}$; i.e., its domain is some subset of $\mathbf{N}$, and let A be a set of natural numbers. The function φ is *Turing reducible* to A if φ is (Turing) computable, relative to A. *See* relative computability. The notation $\varphi \leq_T A$ means that φ is Turing reducible to A. If B is a set of natural numbers, then B is Turing reducible to A ($B \leq_T A$) if its characteristic function χ_B is Turing reducible to A.

For example, given any set A of natural numbers, $\overline{A} \leq_T A$ where $\overline{A}$ is the complement of A in $\mathbf{N}$. If B is any computably enumerable (recursively enumerable) set and K is the halting set $\{e : \varphi_e(e) \text{ is defined}\}$, where φ_e is the partial recursive function with Gödel number e, then $B \leq_T K$.

twin primes   Two odd prime numbers p and q so that $q = p + 2$. For example, 3 and 5 are *twin primes,* as are 5 and 7, 11 and 13, 17 and 19, and 29 and 31. Twin primes with over 3300 digits have been discovered, but it is unknown whether or not there are infinitely twin prime pairs. The triple $(3, 5, 7)$ forms the only "prime triplet" since at least one of any triple of the form $(n, n + 2, n + 4)$ must be divisible by 3.

Tychonoff Fixed-Point Theorem   Suppose X is a locally convex linear topological space and $C \subseteq X$ is compact and convex. Then any continuous function $f : C \to C$ has a fixed point. That is, there is a $c \in C$ with $f(c) = c$. Any normed vector space can be made into a locally convex linear topological space by using the metric topology generated by the norm: $d(x, y) = \|x - y\|$.

Tychonoff space   *See* completely regular topological space.

Tychonoff Theorem   The product of any number of compact topological spaces is compact in the product topology. For example, since the unit interval $[0, 1]$ is compact, any cube $[0, 1]^\kappa$ is also compact. It is this theorem that makes the product (Tychonoff) topology important. The *Tychonoff Theorem* is equivalent to the Axiom of Choice.

Tychonoff topology   *See* product topology.

type   A *type* of a theory T is any set of formulas that is realized in some model of T. That is, if T is a (possibly empty) theory in the language L, then a set of L-formulas $\Phi(\bar{x})$ is an n-type of T if $\bar{x} = \{x_1, \ldots, x_n\}$ contains all free variables occurring in the formulas of Φ, and there is a model A of T and an n-tuple $\bar{a}$ of elements of A such that $A \models \phi(\bar{a})$ for every $\phi(\bar{x})$ in $\Phi(\bar{x})$.

Some authors require types to be complete, meaning they are maximally consistent.

U

ultrafilter   A subset $\mathcal{U}$ of a Boolean algebra B, which is a filter, not properly contained in any other filter on B. As a filter, $\mathcal{U}$ must be nonempty, closed under $\wedge$, not contain 0, and be closed upwards: for all $u \in \mathcal{U}$ and $b \in B$, if $u \leq b$ then $b \in \mathcal{U}$. The maximality condition is equivalent to requiring that for all $b \in B$, either $b \in \mathcal{U}$ or $\neg b \in \mathcal{U}$.

Any filter can be extended to an *ultrafilter*, and, using a weak form of the Axiom of Choice, any subset of a Boolean algebra with the finite intersection property can be extended to an ultrafilter.

ultrapower   An *ultrapower* of an L-structure A is a reduced product $\prod_{\mathcal{U}} A$, where $\mathcal{U}$ is an ultrafilter over the index set I. The reduced product is formed by declaring, for x and y in the Cartesian product $\prod_I A$, that $x \equiv_{\mathcal{U}} y$ if and only if the set of coordinates where x and y agree is in the ultrafilter $\mathcal{U}$:

$$\{i \in I : x(i) = y(i)\} \in \mathcal{U} \, .$$

The reduced product $\prod_{\mathcal{U}} A$ is then the set of all equivalence classes under $\equiv_{\mathcal{U}}$.

The fundamental property of ultrapowers is that, for any L-sentence ϕ, $\prod_{\mathcal{U}} A \models \phi$ if and only if $\{i \in I : A \models \phi\} \in \mathcal{U}$. But because $\mathcal{U}$ is an ultrafilter, $\emptyset \notin \mathcal{U}$ and $I \in \mathcal{U}$, and so, the ultrapower models ϕ if and only if the original structure A models ϕ. Thus, $\prod_{\mathcal{U}} A \equiv A$.

See also ultraproduct.

ultraproduct   An *ultraproduct* of a set of L-structures $\{A_i : i \in I\}$ is a reduced product $\prod_{\mathcal{U}} A_i$, where $\mathcal{U}$ is an ultrafilter over the index set I. *See* ultrapower.

The fundamental property of ultraproducts is that for any L-sentence ϕ, $\prod_{\mathcal{U}} A_i \models \phi$ if and only if $\{i \in I : A_i \models \phi\} \in \mathcal{U}$.

umbilical point   Let M be a surface in $\mathbf{R}^3$, and let $k_1 \geq k_2$ be the principal curvature functions. *See* principal curvature. An *umbilical point* is a point where $k_1 = k_2$. On the complement of the set of umbilical points, the principal curves form a pair of orthogonal fields of curves on the surface; the umbilical points are the places where these fields become singular.

unbounded set   A set of ordinals $C \subseteq \kappa$ such that, for any $\alpha < \kappa$, there is a β with $\alpha \leq \beta < \kappa$ and $\beta \in C$. *See also* closed set, stationary set.

uncountable   A set that is infinite but not denumerable. For example, $\mathbf{R}$ and $\mathbf{C}$ are *uncountable* sets.

undecidable   A set of objects of some sort, which it is not decidable. *See* decidable.

uniformly continuous function   A function $f : \mathbf{R} \to \mathbf{R}$ such that, for any $\epsilon > 0$, there is a $\delta > 0$ such that for x and x' in $\mathbf{R}$, $|f(x) - f(x')| < \epsilon$ whenever $|x - x'| < \delta$. Any continuous $f : [a, b] \to \mathbf{R}$ is uniformly continuous.

More generally, a function f from one metric space (X, d_X) to another (Y, d_Y) is uniformly continuous if for any $\epsilon > 0$, there is a $\delta > 0$ such that, for all x and x' in X, $d_Y(f(x), f(x')) < \epsilon$ whenever $d_X(x, x') < \delta$. If X is compact, then any continuous $f : X \to Y$ is uniformly continuous.

Further generalization of the notion is possible in a uniform space. *See* uniform space.

uniform space   A set X with the topology induced by a uniformity $\mathcal{U}$. Informally, a uniformity is a way of capturing closeness in a topological space without a metric; that is, it provides a generalization of a metric. Formally, a nonempty collection $\mathcal{U}$ of subsets of $X \times X$ is a uniformity if it satisfies the following conditions:

(i.) for all $U \in \mathcal{U}$, $\Delta \subseteq U$, where $\Delta = \{(x, x) : x \in X\}$ is the diagonal of X;

(ii.) for all $U \in \mathcal{U}$, $U^{-1} \in \mathcal{U}$, where $U^{-1} = \{(y, x) : (x, y) \in U\}$;

(iii.) for all U and V in $\mathcal{U}$, $U \cap V \in \mathcal{U}$;

(iv.) for each $U \in \mathcal{U}$ there is a $V \in \mathcal{U}$ with $V \circ V \subseteq U$, where

$$V \circ V =$$

$$\{(x, z) : \exists y \in X \ (x, y) \in V \text{ and } (y, z) \in V\} \, ;$$

1-58488-050-3/01/\$0.00+\$.50
© 2001 by CRC Press LLC

and

(v.) for all $U \in \mathcal{U}$, if $U \subseteq V$, then $V \in \mathcal{U}$. The idea is that x and y will be considered U-close to each other if $(x, y) \in U$. Then, for example, condition (i.) states that x is always U-close to itself.

A uniformity $\mathcal{U}$ generates a topology on X (the uniform topology) by considering the sets $U[x] = \{y : (x, y) \in U\}$ as basic open sets for each $U \in \mathcal{U}$ and $x \in X$.

uniform topology   (1) *See* uniform space.

(2) The *uniform topology* on $\mathbf{R}^\alpha$ is the topology induced by the bounded sup metric

$$\delta(\overline{x}, \overline{y}) = \sup\{\min\{|x_\beta - y_\beta|, 1\} : \beta < \alpha\}.$$

This topology is the same as the product topology if α is finite; if α is infinite, the uniform topology refines the product topology.

union   (1) The *union of any set* X, denoted by $\cup X$, is the set whose elements are the members of the members of X. That is, $a \in \cup X$ if and only if there exists $S \in X$ such that $a \in S$. For example, $\cup\{(0, k) : k \in \mathbf{Z}\} = \mathbf{R}^+$. If X is an indexed family of sets $\{S_\alpha : \alpha \in I\}$, where I is some index set, the union of X is often denoted by $\bigcup_{\alpha \in I} S_\alpha$.

(2) The *union of sets* A and B, denoted by $A \cup B$, is the set of all elements that belong to at least one of A and B. This is a special case of the previous definition, as $A \cup B = \cup\{A, B\}$. For example, $\{3, 10\} \cup \{3, 5\} = \{3, 10, 5\}$ and $\mathbf{N} \cup \mathbf{R} = \mathbf{R}$. *See also* Axiom of Union.

unit function   The arithmetic function, denoted u, which returns the value 1 for all positive integers, i.e., $u(n) = 1$ for all integers $n \geq 1$. (*See* arithmetic function.) It is completely (and strongly) multiplicative.

universal bundle   A bundle $EG \longrightarrow BG$ with fiber G is a *universal bundle* with structure group G if EG is contractible and every G bundle over X is the equivalent to the bundle formed by the pullback of $EG \longrightarrow BG$ along some map $X \longrightarrow BG$.

Example: The universal real line bundle is $EO(1) \longrightarrow BO(1)$ equivalent to the covering of $BO(1) = \mathbf{RP}^\infty$ (infinite dimensional real projective space) by S^∞, the union over all n of spheres S^n, under the action of $\mathbf{Z}/2 = O(1)$.

universal element   If $\mathcal{C}$ is any category, $\mathcal{S}$ is the category of sets and functions, and $F : \mathcal{C} \to \mathcal{S}$ is a functor, a *universal element* of F is a pair (A, B), where A is an object of $\mathcal{C}$ and $B \in F(A)$, such that for every pair (A', B'), where $B' \in F(A')$, there exists a unique morphism $f : A \to A'$ of $\mathcal{C}$ with $(F(f))(B) = B'$.

universal mapping property   The notion of a *universal mapping property* is not a rigorously defined one, as many variations exist. A common pattern that appears in many instances can be described as follows. A triple (p, A, A'), where A and A' are objects of a category $\mathcal{C}$ and $p : A \to A'$ is a morphism of $\mathcal{C}$, has a universal mapping property if, for every morphism $f : X \to A$ of $\mathcal{C}$, there exists a unique morphism $f' : X \to A'$ of $\mathcal{C}$ such that $f' = p \circ f$. In most cases, a universal mapping property is used to define a new object. A standard example of defining a tuple having a universal mapping property is the product of objects in a category. *See* product of objects.

universal quantifier   *See* quantifier.

universal sentence   A sentence σ of a first-order language $\mathcal{L}$ which has the form $\forall v_1 \ldots \forall v_n \alpha$, where α is quantifier-free, for some $n \geq 0$.

universe of sets   The collection of all sets. In Zermelo-Fraenkel set theory (ZFC), the *universe of sets,* usually denoted by V, can be expressed by the abbreviation $V = \bigcup_\alpha V_\alpha$, where each V_α is a set from the cumulative hierarchy. It is important to note that this union does not define a set in ZFC, rather, the above equation is simply an abbreviation for the following statement which is provable in ZFC: $(\forall x)(\exists \alpha)\ x \in V_\alpha$. *See also* cumulative hierarchy.

unordered pair   A set with exactly two elements. For example, $\{3, -5\}$ is an unordered pair. *Compare with* ordered pair.

upper limit topology   *See* Sorgenfrey line.

Urysohn's Lemma   For any two disjoint closed subsets A and B of a normal topological space X, there is a continuous $f : X \to [0, 1]$ such that $f(a) = 0$ for every $a \in A$ and $f(b) = 1$ for every $b \in B$. That is, normality implies disjoint closed sets may be separated by continuous functions. The converse is easier: if f is continuous and separates A and B, then $f^{-1}([0, \frac{1}{2}))$ and $f^{-1}((\frac{1}{2}, 1])$ are disjoint open sets containing A and B, respectively. Thus, normality is equivalent to separation by continuous functions for Hausdorff spaces.

Urysohn's Lemma is a vital part of the proofs of Tietze's Extension Theorem and Urysohn's Metrization Theorem.

Urysohn's Metrization Theorem   Any regular, second countable topological space is metrizable. In other words, if X is regular and has a countable basis, then there is a metric that induces the topology on X. The proof relies on Urysohn's Lemma and imbeds X in the cube $[0, 1]^{\omega}$, which is also separable. *See also* Urysohn's Lemma.

V

valid   Let $\mathcal{L}$ be a first order language and let α be a well-formed formula of $\mathcal{L}$. If, for every structure $\mathcal{A}$ for $\mathcal{L}$ and for every $s : V \rightarrow A$, $\mathcal{A}$ satisfies α with s, then α is *valid* or is a *validity*. (Here, V is the set of variables of $\mathcal{L}$ and A is the universe of $\mathcal{A}$.)

As an example, let $\mathcal{L}$ be the language of equality, $=$. The formula

$$(v_1 = v_2 \wedge v_2 = v_3) \rightarrow v_1 = v_3$$

is valid.

validity   *See* valid.

Venn diagram   A schematic device used to verify relations among sets contained within a universal set U.

The universal set U may be represented by a closed figure such as a rectangle. A set $A \subset U$ is then represented by the interior of some closed region within U, while the statement $x \in A$ is indicated as a point within the region A. The relation $A \subset B$ is depicted by placing the region representing A within that of B.

The union $A \cup B$ of two sets may be represented by shading the combined regions including both A and B. The intersection $A \cap B$ is indicated by shading the overlapping portions of the regions A and B and the complement of A or A' is indicated by shading the region within U which is outside A.

The relation $(A \cup B)' = A' \cap B'$ is shown in the figure. The top diagram indicates by shading the set $(A \cup B)'$ and the bottom diagram indicates the common elements of A' and B'.

von Mangoldt function   *See* Mangoldt function.

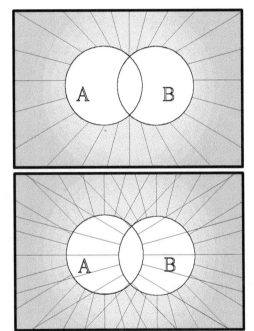

Top: $(A \cup B)'$. *Bottom:* $A' \cap B'$

1-58488-050-3/01/$0.00+$.50

Wang exact sequence   Let $F \longrightarrow E \longrightarrow S^n$ be a fiber bundle with $n \geq 2$ and F path connected. Then there is a long exact sequence

$$\cdots \longrightarrow H^k(E) \longrightarrow H^k(F) \longrightarrow H^{k-n+1}(F)$$

$$\longrightarrow H^{k+1}(E) \longrightarrow \cdots$$

called the *Wang exact sequence*.

This sequence is derived from the spectral sequence for the fiber bundle, which in this case has only one non-trivial differential. There is an analogous sequence for homology. One can use the Wang sequence to compute the homology of the based loop space of a sphere.

wedge   The one-point union of two spaces; in other words, the wedge product of two spaces is formed from their disjoint union by identifying one chosen point in the first space with a chosen point in the second. In the category of pointed spaces (spaces together with a base point), the chosen point is the base point. For example, the *wedge* of two circles is a figure eight.

well-formed formula   In propositional (sentential) logic, a *well-formed formula* (or wff) satisfies the following inductive definition.

(i.) If A is a sentence symbol, then A is a wff.

(ii.) If α and β are wffs, then so are $(\neg \alpha)$, $(\alpha \wedge \beta)$, $(\alpha \vee \beta)$, $(\alpha \rightarrow \beta)$, and $(\alpha \leftrightarrow \beta)$.

(iii.) The set of well-formed formulas is generated by rules (i.) and (ii.).

For example, if A, B, and C are sentence symbols, then $((A \wedge B) \vee C)$ is a wff, while $A\wedge$ is not a wff. Informally, the parentheses used in defining wffs are often omitted when doing so does not affect the readability of the formula; in particular, it is always assumed that $\neg$, $\wedge$, and $\vee$ apply to as little as possible. For example, if A, B, and C are sentence symbols, then $\neg A \wedge B \rightarrow C$ means $(((\neg A) \wedge B) \rightarrow C)$.

In first order logic, with a given first order language $\mathcal{L}$, the set of wffs of $\mathcal{L}$ is defined inductively.

(i.) If α is an atomic formula, then α is a wff.

(ii.) If α and β are wffs, then so are $(\neg \alpha)$ and $(\alpha \rightarrow \beta)$.

(iii.) If α is a wff and v is a variable, then $\forall v \alpha$ is a wff.

(iv.) The set of well-formed formulas is generated by rules (i.), (ii.), and (iii.).

Since $\{\neg, \rightarrow\}$ is a complete set of logical connectives, it is possible to use the other connectives informally in well-formed formulas as abbreviations for formulas in the actual formal language $\mathcal{L}$. In particular, if α and β are well-formed formulas of $\mathcal{L}$, then

(i.) $(\alpha \vee \beta)$ abbreviates $((\neg \alpha) \rightarrow \beta)$.

(ii.) $(\alpha \wedge \beta)$ abbreviates $(\neg(\alpha \rightarrow (\neg \beta)))$.

(iii.) $(\alpha \leftrightarrow \beta)$ abbreviates $((\alpha \rightarrow \beta) \wedge (\beta \rightarrow \alpha))$.

Informally, the parentheses used in defining wffs are often omitted when doing so does not affect the readability of the formula, or even added when doing so aids the readability of the formula. It is always assumed that $\forall$ applies to as little as possible. For example, $\forall v \alpha \rightarrow \beta$ means $(\forall v \alpha \rightarrow \beta)$, rather than $\forall v(\alpha \rightarrow \beta)$.

For example, in the language of elementary number theory (*see* first order language), $\forall v_1 (< (v_1, S(v_1)))$ is a well-formed formula, although $< (v_1, S(v_1))$ is usually informally written as $v_1 < S(v_1)$.

well-founded relation   A partial ordering R, on a set S, such that every nonempty subset of S has an R-minimal element. For example, the relation "m divides n", on the set of natural numbers, is well-founded; the relation $\leq$ on the set of real numbers is not well founded.

well-founded set   A set X on which the membership relation is well founded. That is, any nonempty subset of X contains an ϵ-minimal element. A *well-founded set* cannot contain itself as a member.

well-ordered set   A pair $(S, \leq)$ such that $\leq$ is a well-ordering of S. For example, $(\mathbf{N}, \leq)$ is a *well-ordered set*. Also called *woset*.

well-ordering   A linear ordering $\leq$ of some set S such that every nonempty subset of S has a minimum element. For example, the usual linear ordering $\leq$ for numbers is a *well-ordering* of $\mathbf{N}$ but it is not a well-ordering of $\mathbf{R}$.

Well-Ordering Theorem   Every set can be well ordered; i.e., for every set there exists an ordering on that set which is a well-ordering. *See* well-ordering. The *Well-Ordering Theorem* is equivalent to the Axiom of Choice. *See* Axiom of Choice. Consequently, the Well-Ordering Theorem is independent of the axioms of ZF (Zermelo-Fraenkel set theory); that is, it can neither be proved nor disproved from ZF.

Whitney sum   The sum of two vector bundles over a manifold, formed by taking the direct sum of the vector spaces over each point. The Möbius band M can be thought of as a vector bundle over the circle (since the unit interval $(0, 1)$ is homeomorphic to $\mathbf{R}$). This vector bundle is distinct from the trivial bundle $E = \mathbf{R}^1 \times S^1$, but both *Whitney sums* $E \oplus E$ and $M \oplus M$ are equivalent to the trivial bundle $\mathbf{R}^2 \times S^1$.

whole number   A non-negative integer.

woset   *See* well-ordered set.

Z

Zermelo hierarchy   *See* cumulative hierarchy.

Zermelo set theory   Zermelo-Fraenkel set theory without the Axiom of Replacement. Abbreviated by the letter Z. *See* Zermelo-Fraenkel set theory.

Zermelo-Fraenkel set theory   The formal theory whose axioms are: the Axiom of Extensionality, the Axiom of Regularity, the Axiom of Pairing, the Axiom of Separation, the Axiom of Union, the Axiom of Power Set, the Axiom of Infinity, the Axiom of Replacement, and the Axiom of Choice. This axiomatic theory is often abbreviated as ZFC (the letter C is for the Axiom of Choice).

zero   *Symbol:* 0
(1) A symbol representing the absence of quantity.
(2) The additive identity of an Abelian group A. The element denoted as $0 \in A$ which has the property that $0 + a = a + 0 = a$ for every element $a \in A$.

zero object   An object A of a category C that is both terminal and initial is a *zero object* of C. Such an object is usually denoted by 0 or $*$, and is also called a *null object* of the category. For example, in the category of Abelian groups and group homomorphisms, $(\{0\}, +)$ is a zero object. Any two zero objects are isomorphic.

zero section   A map $M \longrightarrow E$ of a vector bundle $E \longrightarrow M$ over a manifold M, which takes each point m in M to the zero in the vector space which is the fiber over m. That this map is well defined follows from the definition of vector bundle.
Example: For any trivial bundle $M \times \mathbf{R}^n$, $M \times \{0\}$ is the *zero section*.
The term zero section can also refer to the image of the section map.

ZF   Zermelo-Fraenkel set theory without the Axiom of Choice. *See* Zermelo-Fraenkel set theory.

ZFC   *See* Zermelo-Fraenkel set theory.

Zorn's Lemma   If $(\mathcal{P}, \leq)$ is a nonempty partial order in which every chain has an upper bound, then $\mathcal{P}$ has a maximal element. In other words, if for every linearly ordered $C \subseteq \mathcal{P}$ there is a $p_c \in \mathcal{P}$ such that $q \leq p_c$ for all $q \in C$, then there is one $p \in \mathcal{P}$ such that $q \leq p$ for all $q \in \mathcal{P}$. *Zorn's Lemma* is equivalent to the Axiom of Choice.